Sheffield Hallam University
Learning and IT Services
Adsetts Centre City Campus
Sheffield

D1583461

SHEFFIELD HALLAM UNIVERSITY
LEARNING CENTRE
WITHDRAWN FROM STOCK

ONE WEEK LOAN

22 NOV 2013

Developing Intelligent Agent Systems

Wiley Series in Agent Technology

Series Editor: Michael Wooldridge, *Liverpool University, UK*

The 'Wiley Series in Agent Technology' is a series of comprehensive practical guides and cutting-edge research titles on new developments in agent technologies. The series focuses on all aspects of developing agent-based applications, drawing from the Internet, telecommunications, and Artificial Intelligence communities with a strong applications/technologies focus.

The books will provide timely, accurate and reliable information about the state of the art to researchers and developers in the Telecommunications and Computing sectors.

Titles in the series:

Padgham/Winikoff: Developing Intelligent Agent Systems 0470861207   (June 2004)
Pitt (ed.): Open Agent Societies   047148668X (August 2004)

Developing Intelligent Agent Systems
A practical guide

Lin Padgham & Michael Winikoff
RMIT University, Melbourne, Australia

John Wiley & Sons, Ltd

Copyright 2004   John Wiley & Sons Ltd, The Atrium, Southern Gate, Chichester,
West Sussex PO19 8SQ, England

Telephone (+44) 1243 779777

Email (for orders and customer service enquiries): cs-books@wiley.co.uk
Visit our Home Page on www.wileyeurope.com or www.wiley.com

Reprinted December 2005

All Rights Reserved. No part of this publication may be reproduced, stored in a retrieval system or transmitted in any form or by any means, electronic, mechanical, photocopying, recording, scanning or otherwise, except under the terms of the Copyright, Designs and Patents Act 1988 or under the terms of a licence issued by the Copyright Licensing Agency Ltd, 90 Tottenham Court Road, London W1T 4LP, UK, without the permission in writing of the Publisher. Requests to the Publisher should be addressed to the Permissions Department, John Wiley & Sons Ltd, The Atrium, Southern Gate, Chichester, West Sussex PO19 8SQ, England, or emailed to permreq@wiley.co.uk, or faxed to (+44) 1243 770620.

This publication is designed to provide accurate and authoritative information in regard to the subject matter covered. It is sold on the understanding that the Publisher is not engaged in rendering professional services. If professional advice or other expert assistance is required, the services of a competent professional should be sought.

Other Wiley Editorial Offices

John Wiley & Sons Inc., 111 River Street, Hoboken, NJ 07030, USA

Jossey-Bass, 989 Market Street, San Francisco, CA 94103-1741, USA

Wiley-VCH Verlag GmbH, Boschstr. 12, D-69469 Weinheim, Germany

John Wiley & Sons Australia Ltd, 33 Park Road, Milton, Queensland 4064, Australia

John Wiley & Sons (Asia) Pte Ltd, 2 Clementi Loop #02-01, Jin Xing Distripark, Singapore 129809

John Wiley & Sons Canada Ltd, 22 Worcester Road, Etobicoke, Ontario, Canada M9W 1L1

Wiley also publishes its books in a variety of electronic formats. Some content that appears in print may not be available in electronic books.

British Library Cataloguing in Publication Data

A catalogue record for this book is available from the British Library

ISBN 10: 0-470-86120-7 (HB)
ISBN 13: 978-0-470-86120-2 (HB)

Produced from LaTeX files supplied by the author, typeset by Laserwords Private Limited, Chennai, India
Printed and bound in Great Britain by Antony Rowe Ltd, Chippenham, Wiltshire
This book is printed on acid-free paper responsibly manufactured from sustainable forestry
in which at least two trees are planted for each one used for paper production.

Contents

Foreword from the Series Editor

As the concepts and technologies associated with intelligent software agents make their transition from the research lab to the desk of the IT practitioner, issues such as reliable analysis and design methodologies come increasingly to the fore. If agent technology is to mature into a successful and widely used approach to software development, then it is of critical importance that methodologies are developed, which are accessible to students and IT professionals alike, enabling them to deploy this new and promising technology to full effect. Although several such methodologies have been tentatively proposed, the PROMETHEUS methodology set out in this book is arguably the most mature.

PROMETHEUS is a general purpose methodology for the development of software agent systems, in that it is not tied to any specific model of agency or software platform. The authors do an excellent job of describing the models and methods associated with PROMETHEUS and they show how these can be used to analyse and design multiagent systems by means of a detailed running example. Associated with the methodology, the authors have developed a freely available software design tool (PDT), which represents the state-of-the-art in multiagent systems development tools. All in all, this book represents a valuable contribution, not just for those with an interest in the ongoing debate about development methods for multiagent systems but also for those who simply want an answer to the question: How do I actually do it?

Mike Wooldridge, Liverpool, June 2004

Preface

Intelligent software agents[1] are a powerful technology that is attracting considerable (and growing!) interest.

While there are books that cover research areas on agents or survey the field (including the excellent book by Michael Wooldridge (Wooldridge 2002)), there is no book that is aimed at an industrial software developer that answers not only the questions 'what are agents?' and 'why are they useful?' but also the crucial question 'how do I design and build intelligent software agents?'.

Our book aims to provide a practical introduction to building intelligent agent systems. It covers everything a practitioner needs to know to build multi-agent systems of intelligent agents. It includes an introduction to the notion of agents, a description of the concepts involved, and a software engineering methodology covering specification, analysis, design and implementation of agent systems.

The core of the book is the *Prometheus* methodology for designing multi-agent systems. The methodology was developed over the past six or seven years in collaboration with Agent Oriented Software[2], a company that markets the agent development platform, JACK^{TM} as well as agent solutions. The methodology has been used internally at Agent Oriented Software and has also been taught at industry workshops and within university courses. It has proven effective in assisting students and practitioners to develop and document their design and is now at a sufficient level of maturity that support tools have been developed.

Our goal in developing Prometheus was to have a process with associated deliverables that could be taught to industry practitioners and undergraduate students who do not have a background in agents and which they could then use to develop intelligent agent systems. Our evidence that we have achieved this is, at this stage, still anecdotal; however, the indications are that Prometheus *is* usable by nonexperts and that they find it useful.

We do *not* believe that Prometheus is complete and perfect, nor that it is a perfect fit as-is for all applications and all users. However, we do believe that it is usable and in our experience it is much better than not having a methodology. Like most methodologies, Prometheus is intended to be interpreted as a set of guidelines and you should use your common sense and take what is useful, adapting the methodology as needed to suit your needs.

Although we do not believe that Prometheus is perfect, it *is* general purpose in the sense of not being specific to BDI (Belief-Desire-Intention) agents. Only the later part of

---

[1]This is shortened to 'agents' in the remainder of this book.

[2]*http://www.agent-software.com*

the detailed design phase (Chapter 9) makes assumptions about particular types of agent platforms. The assumptions made are fairly general and correspond to a class of agent platforms that have hierarchical plans with triggers, and a description for each plan that indicates the context in which it is applicable.

Electronic Bookstore: Case study

We will be using an example of an electronic bookstore to illustrate the design process throughout the book. To enable easy following of the example, we will enclose all of these examples in a framed box (like this one), which may extend over page breaks, in which case the bottom and top of the frame on the adjacent pages will be missing. In addition, the collected details of the example can be found in Appendix A.

Electronic resources, including the forms in Appendix B and the Prometheus Design Tool (PDT), can be found at

http://www.cs.rmit.edu.au/agents/prometheus

Audience

This book is aimed at industrial software developers and at undergraduate students. It assumes knowledge of basic software engineering but does not require knowledge of Artificial Intelligence or of mathematics. Familiarity with Java will help in reading the examples in Chapter 10.

Tool Support

We believe that tool support is, if not essential, incredibly useful in developing large designs and in helping to keep them consistent. Thus, we have developed a prototype tool, the Prometheus Design Tool (PDT). This tool supports the process described in this book, of system specification, architectural design and detailed design. The detailed design produced by PDT can be straightforwardly converted to the JACK Development Environment (JDE), and consequently to JACK code. A similar approach could be used to develop plug-ins that enable PDT to produce skeleton code for a range of agent programming platforms.

Acknowledgements

This book has evolved out of material originally developed with Agent Oriented Software (AOS) and refined over the course of a number of years of teaching an Agent Oriented Programming and Design course and supervising students in developing multi-agent systems. We thank our colleagues at AOS, especially Andrew Lucas and Ralph Rönnquist, as well as the many students who took our courses and provided valuable feedback.

We acknowledge the support of the Australian Research Council (ARC) under grant CO0106934[3], and its continuation, grant LP0453486[4].

The Prometheus Design Tool was initially developed by Anna Edberg and Christian Andersson. Further development has been done by Claire Hennekam and Jason Khallouf.

We especially thank Ian Mathieson for his careful reading and commenting – this book has benefited considerably from his detailed comments.

Lin Padgham & Michael Winikoff
June 2004
Melbourne, AUSTRALIA

---

[3] *Simplifying the Development of Agent-Oriented Systems*, ARC SPIRT Grant, 2001-2003.
[4] *Advanced Software Engineering Support for Intelligent Agent Systems*, ARC Linkage Grant, 2004-2006.

1

Agents and Multi-Agent Systems

This book is about designing and implementing intelligent agent systems. We therefore begin by answering the obvious first question, namely, '*What* is an agent?'. We answer this question by discussing the properties that characterize an intelligent agent, and contrast agents with objects. The usual second question is '*Why* should I bother with agents?'. We answer this question by arguing that agents are a natural progression from objects that provide a better abstraction and improved encapsulation, and also, perhaps more convincingly, by looking at applications of agent technology.

The remaining chapters of this book are dedicated to answering the third question '*How* do I develop agents and agent systems?'.

1.1 WHAT IS AN INTELLIGENT AGENT?

As is to be expected from a fairly young area of research, there is not yet a universal consensus on the definition of an agent. However, the Wooldridge and Jennings definition (see below) is increasingly adopted, and it is probably fair to say that most researchers in the field, when asked to provide their definition, will mention various properties drawn from those we discuss below.

> ☞ DEFINITION: The following definition is from (Wooldridge 2002), which in turn is adapted from (Wooldridge and Jennings 1995):
>
> > 'An *agent* is a computer system that is *situated* in some *environment*, and that is capable of *autonomous action* in this environment in order to meet its design objectives'.
>
> Wooldridge distinguishes between an agent and an *intelligent* agent, which is further required to be *reactive*, *proactive* and *social* (Wooldridge 2002, page 23).

Developing Intelligent Agent Systems L. Padgham & M. Winikoff
© 2004 John Wiley & Sons, Ltd ISBN: 0-470-86120-7 (HB)

Let us first note that we are talking about *software* agents. Whenever we (or any other researcher in the field) say 'agent', we really mean 'software agent'. The typical dictionary definition of agent as 'an entity having the authority to act on behalf of another' (e.g. a real estate agent) is *not* what we mean[1].

Two basic properties of software agents are that they are **autonomous** and that they are **situated** in an environment. The first property, being *autonomous*, means that agents are independent and make their own decisions. This is one of the properties that distinguishes agents from objects. When we consider a system consisting of a number of agents, then a consequence of the agents being autonomous is that the system tends to be decentralized (we shall return to this in the next section).

The second property (*situatedness*) does not constrain the notion of an agent very much since virtually all software can be considered to be situated in an environment. However, where agents differ is the *type* of environments. Agents tend to be used where the environment is challenging; more specifically, typical agent environments[2] are *dynamic, unpredictable* and *unreliable*. These environments are *dynamic* in that they change rapidly. By 'rapidly', we mean that the agent cannot assume that the environment will remain static while it is trying to achieve a goal. These environments are *unpredictable* in that it is not possible to predict the future states of the environment; often this is because it is not possible for an agent to have perfect and complete information about their environment, and because the environment is being modified in ways beyond the agent's knowledge and influence. Finally, these environments are *unreliable* in that the actions that an agent can perform may fail for reasons that are beyond an agent's control. For example, a robot attempting to lift an item may fail for a wide range of reasons including the item being too heavy.

Agents are often situated in dynamic environments that change rapidly. In particular, this means that an agent must respond to significant changes in its environment. For example, an agent controlling a robot playing soccer can make plans on the basis of the current position of the ball and of other players, but it must be prepared to adapt or abandon its plans should the environment change in a significant way. In other words, agents need to be **reactive**, responding in a timely manner to changes in their environment.

Another key property of agents is that they pursue goals over time, that is, they are **proactive**. One property of goals is that they are *persistent*; this is useful in that it makes agents more robust: an agent will continue to attempt to achieve a goal despite failed attempts.

Although objects can be reactive, and can be seen as having an implicit goal, they are not proactive in the sense of having multiple goals, and of these goals being explicit and persistent. Thus, proactiveness is another property that distinguishes agents from objects.

A key issue in agent architectures is balancing reactiveness and proactiveness. On the one hand, an agent should be reactive, so its plans and actions should be influenced by environmental changes. On the other hand, an agent's plans and actions should be influenced by its goals. The challenge is to balance these two (often conflicting) influences: if the agent is too reactive, then it will be constantly adjusting its plans and not achieve

---

[1] Some software agents may act as agents in this sense as well. For example, a software assistant that buys products or services on behalf of its user.

[2] A more detailed analysis of different properties of environments can be found in (Wooldridge 2002) based on the taxonomy of (Russell and Norvig 1995).

its goals. However, if the agent is not sufficiently reactive, then it will waste time trying to follow plans that are no longer relevant or applicable.

Since failure of actions (and, more generally, of plans) is a possibility in challenging environments, agents must be able to recover from such failures, that is, they must be **robust**. A natural approach to achieving robustness is to be **flexible**. By having a range of ways of achieving a given goal, the agent has alternatives that can be used should a plan fail. These two properties are also distinguishing features of agents as compared to objects.

Finally, agents almost always need to interact with other agents, that is, agents are **social**. This interaction is often at a higher level: instead of just saying that agents exchange messages, agent interaction can be framed in terms of *performatives*[3] such as 'inform', 'request', 'agree', and so on. These have standard semantics that are defined in terms of their effects on an agent's mental state. Agent interaction is often viewed in terms of human interaction types such as negotiation, coordination, cooperation and teamwork.

On the basis of these properties, we use the following definition:

☞ DEFINITION: An **Intelligent Agent** is a piece of software that is

- **Situated** – exists in an environment

- **Autonomous** – independent, not controlled externally

- **Reactive** – responds (in a timely manner!) to changes in its environment

- **Proactive** – persistently pursues goals

- **Flexible** – has multiple ways of achieving goals

- **Robust** – recovers from failure

- **Social** – interacts with other agents.

In addition to these properties, there are a number of other properties that we regard as less central. This does not mean that these are not important, just that they are not important for all agent applications.

In pursuing their goals, we want agents to be **rational**. Part of being rational is that an agent should not do 'dumb' things such as simultaneously committing to two courses of action that conflict. For example, planning to spend money on a holiday at the same time as planning to spend that same money on a car. A detailed analysis of what is meant by 'rational' can be found in the work of Bratman (Bratman 1987). This analysis forms the basis of the Belief-Desire-Intention model for software agents (Rao and Georgeff 1992).

---

[3]This is based on *speech act theory*, which is beyond the scope of this book, and we refer the reader to Chapter 8 of (Wooldridge 2002) for more details.

One definition of agents (*strong agency*) takes these various properties, and also requires that agents are viewed as having *mental attitudes* such as beliefs, goals and intentions. This *intentional stance* (Dennett 1987) has a surprisingly pragmatic justification: as a system becomes more complex, its behaviour can be predicted more reliably by abstracting away from how it achieves its goals and instead reasoning about what are its goals and beliefs. For example, in attempting to ascertain whether a piece of furniture will support a person's weight, we could model the stress and calculate its load-bearing ability, or we could consider its design and reason that the goal of a chair is to be sat upon and so any chair should be able to support a person's weight. The former stance is the 'physical stance', the latter is the 'design stance'. The 'intentional stance' is an extension of this that is applied to active entities.

Although having agents that *learn* from their experiences can be essential for some applications, it can be disastrous for others. Similarly, there are applications in which modelling human emotions can be useful, such as interface agents or computer games, but equally, there are many applications in which this is not relevant.

There is a whole body of work devoted to *mobile* agents (Harrison *et al.* 1995; Kotz and Gray 1999). However, there is surprisingly little overlap between the work on intelligent agents and the work on mobile agents. Mobility is more of a system-level issue, with much work devoted to questions such as 'How can a running program be stopped, moved to another machine and restarted?' and associated issues in security.

1.2 WHY ARE AGENTS USEFUL?

Having described *what* agents are, we now turn to the question of *why* agent technology is useful. It is important to realize that, like other software technologies such as objects, agents are not magic. They are simply an approach to structuring and developing software that offers certain benefits, and that is very well suited to certain types of applications (in fact, one viewpoint considers agents to be an evolutionary step forward from objects (Odell 2002)). In order to understand why agents are useful, we need to understand how the distinctive features of agents translate into properties of software systems that are designed and built using agents. The usefulness of these properties (such as being decentralized) depends on the application, and so it is important to also understand how these software-system properties relate to application types and application areas.

In addition to looking at application types, we try to provide examples of documented applications. Unfortunately, the field of agents is still quite young, so there are not many well-documented applications. However, some applications have been documented in the literature (e.g. Jennings and Wooldridge (1998b)).

Perhaps the single most important advantage of agents is that they reduce coupling. Agents are autonomous, which can be seen as encapsulating invocation (Odell 2002; Parunak 1997). Whereas an object makes available methods that are triggered externally, an agent does not provide any control point to external entities. When a message is sent to an agent, the agent (being autonomous) has control over how it deals with the message.

Coupling is reduced not only by the encapsulation provided by autonomy but also by the robustness, reactiveness and proactiveness of agents. An agent can be relied upon to persist in achieving its goals, trying alternatives that are appropriate to the changing environment. This means that when an agent takes on a goal, the responsibility for achieving that goal rests with that agent. Continuous supervision and checking is not needed. As an analogy, view an object as a reliable employee that lacks initiative and a sense of responsibility; supervising such an employee requires a significant amount of communication. On the other hand, an agent can be viewed as an employee that has a sense of responsibility and shows initiative. Supervising such an employee requires considerably less communication, and hence less coupling.

Reduced coupling can lead to software systems that are more modular, more decentralized and more changeable. This has led to the application of agents as an architectural 'glue' in a range of software applications. In this usage, agents are often used to 'wrap' legacy software. For example, see the list of applications built with Open Agent Architecture (OAA) listed in (Cheyer and Martin 2001).

It has been argued that agents are 'well suited for developing complex distributed systems' (Jennings 2001) since they provide more natural abstraction and decomposition of complex 'nearly-decomposable' systems.

One increasingly important class of systems that exhibit decentralization, complexity and distribution is *open systems*: software systems in which different parts are designed and written by different authors, without there being communication between the different authors. An example is the World Wide Web, where the authors of a web browser and of a web server probably did not ever talk to each other. Not surprisingly, standards play a key role in enabling software that was independently developed to work together. The web is a simple example of an open system since it is essentially concerned with transporting static documents, as opposed to providing services that change the state of servers. Other, more complex, examples of such systems include the semantic web and web services (Hendler 2001; McIlraith *et al.* 2001), and grid computing (Moreau 2002; Moreau *et al.* 2002).

In addition to providing reduced coupling, agents are also clearly applicable in situations in which the environment is challenging (dynamic, unpredictable, unreliable), in which failure is a possibility and in which recovery from failure must be done autonomously. An extreme example of an agent system that was required to deal with such situations was Remote Agent (Muscettola *et al.* 1998), which, in May 1999, was in control of NASA's Deep Space 1 for two days, over 96 500 000 kilometres from the Earth.

Being proactive and reactive makes agents more human-like in the way they deal with problems. This has led to a number of applications in which software agents are used as substitutes for humans in certain limited domains. One application is the use of software agents to substitute for human pilots in military simulations (Tidhar *et al.* 1998). Other, more peaceful, applications include entertainment. The recent computer game *Black & White* used agents, specifically based on the Belief-Desire-Intention (BDI) model that is widely used in the agents community:

> ' . . . To make agents who were psychologically plausible, we took the Belief-Desire-Intention architecture of an agent, fast becoming orthodoxy in the agent programming community, and developed it in a variety of ways . . . '
> – *http://www.gameai.com/blackandwhite.html*

Another area where agents have been applied is in film-making. The recent film *Lord of the Rings: The Two Towers* used a software package called *Massive* to generate the armies of Orcs, Elves and Humans. Each individual character was modelled as an agent.

Other application areas where software agents can provide benefits include Intelligent Assistants (Maes 1994), Electronic Commerce (Luck *et al.* 2003), Manufacturing (Shen and Norrie 1999), and Business process modelling (Jennings *et al.* 2000a,b).

2

Concepts for Building Agents

In the previous chapter, we defined agents as having a number of properties such as being situated, proactive and reactive. In this chapter, we begin to look at how we can design and build software that has these properties. We begin by considering what *concepts* lead to agents having certain properties. For example, in order for an agent to be proactive, it needs to have goals. Thus, the concept of a goal is an important one for designing and building proactive agents.

A software-engineering methodology assumes the existence of a set of concepts that it builds upon. For example, object-oriented notations such as UML (Booch *et al.* 1999) assume certain concepts such as object, class, inheritance, and so on. With agent-oriented methodologies, we also need an appropriate set of underlying concepts, and, not surprisingly, it turns out that the set of concepts is different to the object-oriented set.

The concepts that we describe in this chapter are used by the Prometheus methodology, which is the methodology covered in detail within the later sections of this book. Prometheus has been developed specifically in response to a need for assistance and direction in designing and building agent systems.

Our experience has been that the concepts identified are both necessary and sufficient for building the sort of applications that are appropriately approached using plan-based agents, and that they are simple and can be understood by undergraduate students. These concepts are based on the definition of agents, and in the remainder of this chapter, we explain these concepts and why they are appropriate.

2.1 SITUATED AGENTS: ACTIONS AND PERCEPTS

We began our definition of an agent with the basic property that an agent is software that is situated in an environment. The two concepts that capture the interface between an agent and its environment are the **percepts** from the environment and the **actions** that the agent can perform to affect the environment (Russell and Norvig 1995) (see Figure 2.1).

Developing Intelligent Agent Systems L. Padgham & M. Winikoff
© 2004 John Wiley & Sons, Ltd ISBN: 0-470-86120-7 (HB)

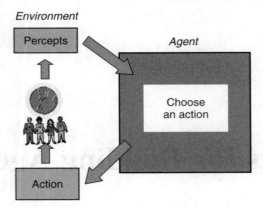

Figure 2.1   Agents are situated. Reproduced from Winikoff *et al.*, (2001), by permission of Springer-Verlag GmbH & Co. KG

A **percept**[1] is an item of information received from the environment by some sensor. For example, a fire-fighting robot may receive information such as the location of a fire and an indication of its intensity. An agent may also obtain information about the environment through sensing actions.

An **action** is something that an agent does, such as move_north or squirt. Agents are situated, and an action is basically an agent's ability to affect its environment. In their simplest form, actions are atomic and instantaneous and either fail or succeed. In the more general case, actions can be *durational* (encompassing behaviours over time) and can produce partial effects; for example, even a failed move_to action may well have changed the agent's location. In addition to external actions that directly affect the agent's environment, we also want to consider *internal actions*. These correspond to an ability that the agent has, which is not structured in terms of plans and goals. Typically, the ability is a piece of code that either already exists or would not benefit from being written using agent concepts, for example, image processing in a vision sub-system.

At a very abstract level, we can view an agent as receiving percepts from an environment, somehow selecting an action to perform, and performing that action. These repeated steps form the execution cycle of an abstract agent. The concepts discussed in the following sections refine the internal execution of the agent. For example, by adding goals, these are able to be used to execute a series of actions over a period of time.

2.2 PROACTIVE AND REACTIVE AGENTS: GOALS AND EVENTS

We want our intelligent agents to be both *proactive* and *reactive*. The agent's proactiveness implies the use of **goals**. A *reactive* agent is one that will change its behaviour in

---

[1] From the Latin 'perceptum', which is the root of the English word 'perceive'.

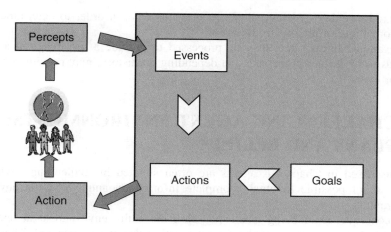

Figure 2.2 Proactive agents achieve goals, reactive agents respond to events. Reproduced from Winikoff *et al.*, (2001), by permission of Springer-Verlag GmbH & Co. KG

response to changes in the environment. An important aspect in decision-making is balancing proactive and reactive aspects. On the one hand we want the agent to stick with its goals by default and on the other hand we want it to take changes in the environment into account. The key to reconciling these aspects, thus making agents suitably reactive, is identifying *significant* changes in the situation. These are **events** (see Figure 2.2).

A **goal** (variously called *task*, *objective*, *aim* or *desire*) is something the agent is working on or towards, for example, extinguish_fire, or rescue_civilian. Often, goals are defined as states of the world that the agent wants to bring about; however, this definition does not allow some types of goals to be expressed such as maintenance goals (e.g. maintain cruising altitude), avoidance goals or safety constraints (e.g. never move the table while the robot is drilling). Goals give the agent its autonomy and proactiveness. An important aspect of proactiveness is the persistence of goals: if a plan for achieving a goal fails, then the agent will consider alternative plans for achieving the goal in question, until it is believed impossible or is no longer relevant.

An **event** is a significant occurrence that the agent should respond to in some way. Events are often extracted from percepts, although they may be generated internally by the agent, for example, on the basis of a clock. An event can trigger new goals, cause changes in information about the environment and/or cause actions to be performed immediately. Actions generated directly by events correspond to *reflexive* actions, executed without deliberation. Events are important in creating reactive agents in that they identify important changes that the agent needs to react to.

Percepts can be seen as particular kinds of events that are generated within the environment. We note that the percept may well have to be interpreted from the raw data available, in order to provide the percept/event that has significance to the agent. Particularly, if the raw data is an image data, it is likely to require significant processing. The first layer would extract features in the image, but even then it is likely to need to be compared to either existing beliefs, or previous data, in order to determine whether there is something of interest to the agent.

For example, bitmap data may be processed to provide a fire image at a location. This needs to be further processed, and depending on history, may provide a `new fire` percept or event.

2.3 CHALLENGING AGENT ENVIRONMENTS: PLANS AND BELIEFS

As we discussed in Chapter 1, agents are often situated in challenging environments where it is not possible to obtain complete information and where the environment changes rapidly.

One consequence of being unable to sense the entire environment at once is that an agent needs to maintain a cache for information that it has received. These are the agent's **beliefs**. A **belief** is some aspect of the agent's knowledge or information about the environment, itself or other agents. For example, an agent might believe there is a fire at X because it saw it recently, even if the agent cannot see the fire now.

Because agents are situated in dynamic environments, it is not realistic for them to use traditional AI planning, that is to assemble plans from actions. Although planning technology and computational speed are improving, planning from action descriptions[2] is still incompatible with real-time decision-making. Instead, agents typically use some kind of 'library of recipes' rather than plan from first principles. This 'library of recipes' is a collection of **plans** that is written by the software developer. A **plan** is a way of realizing a goal; for example, a plan for achieving the goal `extinguish_fire` might specify the three steps: determine a route to the fire, follow the route to the fire and squirt the fire until it has been put out.

Thus, agents in realistic applications that usually have limited computational resources and limited ability to sense their environment need beliefs and plans (see Figure 2.3). Although both of these concepts are 'merely' aids in efficiency, they are not optional. Beliefs are essential since an agent has limited sensory ability and needs to build up its knowledge of the world over time. Plans are also necessary both for computational reasons and for representational reasons. Planning requires that the developer specify each action's preconditions and effects. However, representing the effects of continuous actions operating over time and space in an uncertain world, in sufficient detail for first principles planning, is unrealistic for large applications. Pre-defined plans, however, require only that actions must provide some means for them to be executed.

The notion of plan that we use is fairly general and encompasses a range of plan-based agent platforms including those based on the Belief-Desire-Intention (BDI) model. A plan for achieving a goal provides (a) a function that indicates whether the plan is worth trying in the current situation (i.e. whether it is *applicable*); and (b) a plan body that can be executed in order to (attempt to) achieve the goal. We structure plan bodies in terms of steps that can include the achievement of sub-goals.

Each goal is potentially achievable by a number of different plans. At run time, the agent will select a plan for achieving a given (sub-)goal. If this selected plan fails,

---

[2]These are typically expressed in terms of pre- and postconditions.

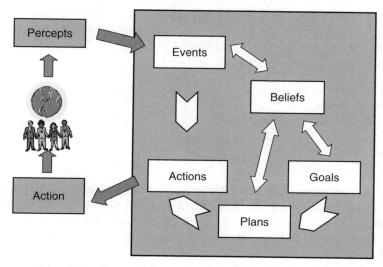

Figure 2.3 Adding plans and beliefs. Reproduced from Winikoff *et al.*, (2001), by permission of Springer-Verlag GmbH & Co. KG

then another plan will be tried. By doing this, agents are flexible, since they can have multiple plans to achieve a given (sub-)goal, and robust, since failure of a plan does not necessarily mean that the goal cannot be achieved.

For example, consider a simplified[3] example of a soccer-playing robot, driven by intelligent agent software. A high-level goal of the robot is to have its team kick goals. The plans available to achieve this may be as follows:

Plan: Kick direct
Goal achieved: Score goal
Applicable if: Robot at ball AND goal unobstructed
Plan body:
1. SUB-GOAL: Face goal
2. ACTION: Kick

Plan: Kick to teammate
Goal achieved: Score goal
Applicable if: Robot at ball AND goal obstructed
Plan body:
1. SUB-GOAL: Identify best teammate to score
2. SUB-GOAL: Face teammate
3. ACTION: Kick

Plan: Get ball
Goal achieved: Score goal

---

[3]These plans are for illustration only. There are a number of complexities in this application that are not addressed by these sample plans.

Applicable if: Robot not at ball AND robot knows ball location
Plan body:
1. SUB-GOAL: Move to *ball location* facing goal
2. SUB-GOAL: Score goal

Plan: Find ball and score
Goal achieved: Score goal
Applicable if: Ball location not known
Plan body:
1. SUB-GOAL: Find ball
2. SUB-GOAL: Score goal

These plans enable the robot to achieve its goal in different ways, depending on its beliefs about the situation it is in. Note that it does not have to *see* the ball to use the plan Get ball, it only has to *believe that it knows* where the ball is. Of course, this plan is a simplification, in that if this belief changes during the execution of the plan, the plan should fail.

As indicated before, the concept of events, as situational information that the agent should *react* to, is also important. Like goals, events are also handled by plans, but the difference between goal and event handling is that once an event is handled, it is gone. Goals, on the other hand, are persistent: if a goal resulted in the execution of a plan that did not achieve the goal, then the agent will continue to try to achieve the goal, if it is able. For example, if the plan above to run and kick failed because the agent no longer knew where the ball was, it would continue to try to achieve its goal by executing the plan Find ball and score. Plans then can either *achieve goals* or *handle events*.

An example of an external event, or percept, to which the robot would react, is seeing the ball at a new location. The following plan could handle this.

Plan: Update ball location
Event handled: Ball seen at X
Applicable if: NOT believed ball at X
Plan body:
1. Update ball location value
2. Update ball location history

2.4 SOCIAL AGENTS

The final attribute of agents according to the definition given in the previous chapter is that they are *social*, that is, agents interact with other agents. There are many forms that this interaction can take, ranging from exchanging messages according to pre-defined protocols to forming teams that work towards a common goal (Cohen and Levesque 1991).

The basic concept used in Prometheus is that of a **message**. A message is a single (one-way) communication. Prometheus does not require that messages are expressed in terms of performatives.

In addition, the concept of a **protocol** is used. A protocol is a definition of the legal interaction patterns (i.e. the sequences of messages that form a conversation). For example, a protocol for placing a book order might require that the message from a customer placing an order must be followed by a request from the sales assistant agent to the customer asking for credit card details. This message can be followed either by a refusal from the customer (in which case the conversation is over) or by credit card details.

Prometheus currently does not provide specific support for designing teams of agents that work towards common goals (Cohen and Levesque 1991), nor does it provide specific and specialized support for designing open systems, though we note that Prometheus does allow for modelling of interaction with a dynamic environment, and that an open system is in fact just that. Software-engineering methodologies for designing open agent systems is still an active area of research and there are no methodologies that are mature. Preliminary work in this area includes the ROADMAP (Juan *et al.* 2002), Nemo (Huget 2002) and RIO (Mathieu *et al.* 2003) methodologies.

2.5 AGENT EXECUTION CYCLE

The concepts of actions, percepts, events, goals, plans and beliefs are related to each other via the **execution cycle** that implements the decision-making of the agent. The execution cycle describes how instances of these concepts interact as an agent executes. For example, how percepts modify beliefs that in turn affect the agent's choice of plans to achieve its goals.

An agent's execution follows a sense-think-act cycle, where the think part of the cycle involves rational decision-making. The cycle can be seen as consisting of the following steps, depicted in Figure 2.4:

1. Events are processed to update beliefs and generate immediate actions.

2. Goals are updated: (i.e. new goals are generated, achieved or impossible goals are dropped, goal priorities are determined).

3. Plans are selected from the plan library for achieving goals or handling events.

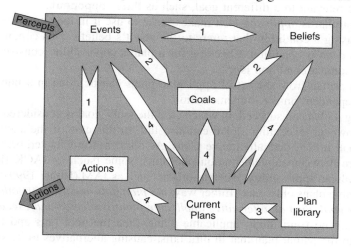

Figure 2.4 Agent execution cycle

4. A plan step is executed in the next plan, yielding new events, (sub)goals, belief changes or actions.

For example, consider a fire-engine robot that receives a percept containing information about a fire at a certain location. Since the agent has no existing knowledge about the fire, this percept results in updating of the agent's beliefs with knowledge of this new fire. This in turn generates a goal to put out the fire, which then leads to a plan being selected and executed.

The execution cycle described is high level and fairly abstract. Of course, the details need to be developed depending on the domain. Extracting relevant information from percepts and updating beliefs appropriately will depend on the application. However, there are generic steps that are part of the standard execution cycle that implements BDI-style decision-making. In the remainder of this section, we focus on describing these generic steps in some detail, as understanding them is important for developing effective systems using this model.

2.5.1 CHOICE OF PLAN TO EXECUTE

The process for handling an event or attempting to achieve a goal follows these steps:

1. Determine the *relevant* plans from the library of plans.

2. Determine the subset of the relevant plans that is *applicable*.

3. Select one of the applicable plans.

4. Execute the selected plan.

A plan is *relevant* if it specifies that it can achieve the goal in question. For example, the Kick direct plan presented earlier is relevant if the goal is to Score goal. It would not be relevant to a different goal, such as Mark opponent.

A relevant plan is *applicable* if it makes sense to use it in the current situation. This is specified using a condition that refers to the agent's beliefs. This condition is known as the plan's *context* condition. Checking for a plan's applicability consists of checking whether the context condition is true.

Selecting a plan from the set of applicable plans can be done in a number of ways and this is dependent on the implementation.

Executing a plan can succeed, in which case the (sub-)goal is considered to have been achieved. However, executing a plan can also fail. In this case, if the agent is trying to achieve a goal, it considers alternative plans. As discussed in Chapter 10, there is some variation here between implementation platforms: some (such as JACK (Busetta *et al.* 1999)) re-evaluate applicability, whereas others (such as JAM (Huber 1999)) compute the set of applicable plans only once. Either way, *if a plan fails and there are other applicable plans for the goal, then an alternative applicable plan is selected and executed.*

If there are no (remaining) applicable plans, then the goal fails and this failure is propagated to the parent plan that in turn fails, causing alternatives to be sought.

Example of agent execution

Let us consider a simple, if slightly artificial, example in which an agent has the following four plans.

Plan: Walk to University (P1)
Goal achieved: Go to University
Applicable if: It is not raining
Plan body: Walk to University

Plan: Train to University (P2)
Goal achieved: Go to University
Applicable if: It is raining
Plan body: SUB-GOAL: Catch Train

Plan: Tram to University (P3)
Goal achieved: Go to University
Applicable if: *true* (always applicable)
Plan body: SUB-GOAL: Catch Tram

Plan: Catch public transport (P4)
Goal achieved: Catch X
Applicable if: *True* (always applicable)
Plan body:
1. Walk to station for X
2. Check timetable
3. Fail if long wait
4. Catch X.

Let us now trace through a sample execution. The agent is trying to achieve the goal Go to University. It begins by determining that P1, P2 and P3 are relevant (but not P4). Assuming that it is currently raining, the agent then determines that P2 and P3 are applicable (but not P1). The agent picks P2 and begins executing it, which results in the sub-goal Catch Train.

The agent is now trying to achieve Catch Train, there is only a single plan that is relevant, and this plan is also applicable. The agent executes this plan, walks to the train station, checks the timetable, and, supposing that the agent has just missed a train, the plan fails because of a long wait.

The agent now considers alternative ways of achieving Catch Train. Since there are no more applicable plans, Catch Train has failed. This means that the parent plan, P2, has failed and so the agent now needs to consider alternative ways of achieving Go to University.

Let us assume that while the agent walks to the train station, it has stopped raining (Melbourne is known for its rather changeable weather!). The agent determines that P1 and P3 are applicable and decides to walk to University (plan P1).

This execution trace illustrates a number of points:

- Although the agent does not generate new plans from scratch, it does combine plans hierarchically.

- The choice of plans to achieve a goal, or sub-goal, takes the current state of the environment into account: the agent chose to walk after it stopped raining.

- By having a number of plans that can be used to achieve a given goal (or react to a given percept), the agent is *flexible*.

- If a plan fails, then the agent will try alternative plans. This makes the agent *robust*.

2.5.2 MANY WAYS TO ACHIEVE A GOAL

Each goal can, in general, have a number of plans that can be used to achieve it. Each plan can have a number of sub-goals that themselves can have multiple applicable plans. This can naturally be depicted in a *goal-plan tree* (see Figure 2.5). The children of each goal are *alternative* ways of achieving that goal (OR) whereas the children of each plan are sub-goals that must all be achieved in order for the plan to succeed (AND).

What is not obvious from the simple example is the large number of alternative ways of achieving a top-level goal that can be generated by a goal-plan tree. If we denote by C the number of plans that are applicable for each goal, by S the number of sub-goals for each plan, and by D the depth of the goal-plan tree, then the number of ways in which the goal at the root of the goal-plan tree can be achieved is

$$C^{((S^D-1)/(S-1))}$$

unless $S = 1$, in which case the number of options is just C^D. The sidebar below gives the derivation of this formula and explains the assumptions made.

As an example, suppose $C = 2$ and $S = 4$ so each nonleaf plan has four sub-goals and each (sub-)goal has two applicable plans. Then the number of possible ways of achieving the top-level goal for a goal-plan tree of depth 3 is $C^{(S^D-1)/(S-1)} = 2^{(4^3-1)/(4-1)} = 2^{(64-1)/3} = 2^{21} = 2\,097\,152$.

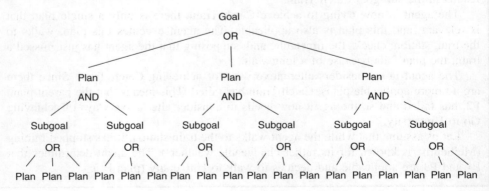

Figure 2.5   Goal-plan tree of depth 2

Thus, we see that this style of programming allows for an enormous number of ways of achieving a goal, without ever needing to program this explicitly. The use of applicability depending on the current environment ensures that only a small and relevant part of this space is actually explored in any given execution. A key to making effective use of this potentially large space of possibilities is to use sub-goals liberally (rather than 'in-line' code), and to separate out different possibilities into alternative plans.

Calculation of number of possibilities

Let C ('C' for 'choice') be the number of plans that are applicable for each goal. Let S be the number of sub-goals per plan, except for the leaves of the goal-plan tree that do not have any sub-goals. Finally, let D be the depth of the tree, measured in terms of the number of goal levels. In the example goal-plan tree above, S and D are 2 and C is 3.

So, the question is this: given a collection of goals and plans, depicted as a goal-plan tree, how many possible ways of achieving the top-level goal are there? Because we are aiming to illustrate the large number of possibilities, we make the simplifying assumption that the tree is uniform in that C is the same for all goals and S the same for all plans.

Let G be a goal that has C applicable plans in which each plan has no sub-goals, then G can be achieved in C possible ways.

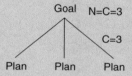

Let P be a plan with S sub-goals in which each sub-goal can be achieved in m possible ways. Since all of the sub-goals must be (separately) achieved, the number of ways that P can be executed is the product $m \times \ldots \times m$, that is, m^S.

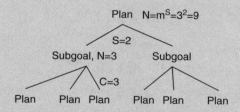

Let G be a goal that has C applicable plans in which each plan can be executed in p possible ways. Since each plan is an alternative, the number of ways that G can be achieved is the sum $p + \ldots + p$, that is, $C \times p$.

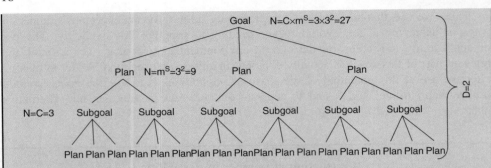

We formalize this by defining $\triangle_G(D)$, the number of ways that a goal at the root of a tree of depth D can be achieved. Clearly, $\triangle_G(1) = C$.

For $\triangle_G(D + 1)$ in which the goal has C child plans $\triangle_G(D + 1) = C \times \triangle_P(D + 1)$, where $\triangle_P(D)$ is the number of ways in which a plan at the root of a tree of depth D can be executed. For $\triangle_P(D + 1)$ in which the plan has S sub-goals, we have that $\triangle_P(D + 1) = \triangle_G(D)^S$. Thus, we have

$$\triangle_G(1) = C$$
$$\triangle_P(D + 1) = \triangle_G(D)^S$$
$$\triangle_G(D + 1) = C \times \triangle_P(D + 1)$$
$$= C \times \triangle_G(D)^S$$

Expanding out this definition, we obtain

$$\triangle_G(1) = C$$
$$\triangle_G(2) = C \times \triangle_G(1)^S$$
$$= C \times C^S$$
$$= C^{S+1}$$
$$\triangle_G(3) = C \times \triangle_G(2)^S$$
$$= C \times C^{(S+1)S}$$
$$= C \times C^{(S+1) \times S}$$
$$= C \times C^{S^2+S}$$
$$= C^{S^2+S+1}$$
$$\triangle_G(4) = C \times \triangle_G(3)^S$$
$$= C \times C^{(S^2+S+1)^S}$$
$$= C \times C^{S^3+S^2+S}$$
$$= C^{S^3+S^2+S+1}$$

Thus, more generally, we have that for a goal-plan tree of depth D the number of options is $C^{S^{(D-1)}+...+S^2+S+1}$.

This can be simplified as follows. Consider the sum $x^{(n-1)} + \ldots + x^2 + x + 1$:

$$\text{Let } \Sigma = x^{n-1} + \ldots + x^2 + x + 1$$

$$x\Sigma = x^n + \ldots + x^2 + x$$

$$(x\Sigma) - \Sigma = x^n - 1$$

$$(x-1)\Sigma = x^n - 1$$

$$\Sigma = \frac{x^n - 1}{x - 1}$$

(if $x = 1$ then $\Sigma = 1^{n-1} + \ldots + 1^1 + 1 = n$).

Therefore, $C^{S^{(D-1)}+...+S^2+S+1}$. can be simplified to $C^{((S^D-1)/(S-1))}$ (unless $S = 1$ in which case the number of options is just C^D).

2.6 SUMMARY

In this chapter, we presented the concepts that we see as being important and useful to designing intelligent agents. These concepts are closely related to the popular *Belief-Desire-Intention* (BDI) model, though with some modifications, which though relatively minor, we consider important for the practical work of building systems.

The BDI model (Georgeff and Rao 1998; Georgeff *et al.* 1999; Rao and Georgeff 1991, 1992) has its basis in philosophy (Bratman 1987) and offers a *logical theory* that defines the mental attitudes of Belief, Desire and Intention; a *system architecture*; a *number of implementations of this architecture* (e.g. PRS (Georgeff and Lansky 1986), JAM (Huber 1999), dMars (AAII 1996), JACK (Busetta *et al.* 1999)); and *applications* demonstrating the viability of the model.

The central concepts in the BDI model are (Georgeff and Rao 1998, page 144)

Beliefs: Information about the environment

Desires/Goals: Objectives to be accomplished[4]

Intentions: The currently chosen course of action

Plans: Means of achieving certain future world states.

In comparison, we have identified the key concepts as being the following:

Actions: Ways the agent can operate on the environment

---

[4]The subtle difference between goals and desires is that goals are required to be consistent, whereas desires may be inconsistent.

Percepts: Relevant information from the environment

Events: Relevant information about a change in situation (percepts are a subset of events)

Goals: Objectives to be accomplished (should be consistent)

Beliefs: Information about the environment (unchanged)

Plans: Means of achieving goals

Messages: Necessary for agents to interact

Protocols: Specifications of interaction 'rules' — usually associated with achieving of goals.

These differ from the BDI model in that we place a greater focus on the situatedness of the agent, and the modelling of the environmental interface via percepts and actions. We also include a multi-agent focus by incorporating messages and protocols. We do not include intentions as it has proved difficult to teach this concept. While the concept of intentions are relevant for the underlying philosophical (or logical) foundation on which the BDI execution cycle is based, it is not necessary for designing and building multi-agent systems.

3

Overview of the Prometheus Methodology

In the previous chapter, we covered the concepts that are used by intelligent software agents. However, there is still a gap – knowing these concepts does not answer the practical question 'How do I build a software system based on intelligent agents?'

The next few chapters provide a detailed answer to this question. What is missing is a *process* that breaks down 'build a software system' into smaller steps that are followed in order to specify, design and build agent-oriented systems.

In addition to high-level steps such as 'specify the system' or even 'identify the system's goals', a usable methodology needs to provide detailed guidelines explaining *how* these steps are carried out. Often, these guidelines are expressed as a collection of heuristics and examples: it is difficult to give hard rules in a general-purpose methodology, and the design decisions often concern trade-offs. Processes and heuristics can help designers identify the decision points and the reasons for making various choices, but cannot make choices for them.

As the process is followed, *design artifacts* are produced – these are used to capture information about the system and its design. For example, in UML, a class diagram captures the classes in a system and their relationships. Design artifacts vary in formality and size – from a weighty tome written in natural English, to diagrams on a white board, to fully formal specifications written in a precisely defined notation.

Although design artifacts could all be specified using a natural language such as English, this is undesirable for a number of reasons: natural languages are unstructured, inherently ambiguous and do not capture certain types of information well. Also, English cannot be easily supported by software tools. Finally, many software engineers do not feel comfortable writing.

As a result, design artifacts are often specified in some formal or, more usually, semi-formal *notation*. A notation can be graphical or textual (including free English text, structured text such as forms, etc.).

Developing Intelligent Agent Systems   L. Padgham & M. Winikoff
© 2004 John Wiley & Sons, Ltd   ISBN: 0-470-86120-7 (HB)

Thus, a methodology should provide a *process* with *detailed guidelines* (including examples and heuristics) and *notations* that are used to describe the design artifacts.

In the remainder of this chapter, we

- discuss why existing (non-agent) methodologies are not appropriate for designing agent systems;

- give a brief overview of the Prometheus methodology;

- provide guidelines regarding the use of Prometheus; and

- briefly relate Prometheus to existing agent-oriented methodologies.

3.1 WHY A NEW METHODOLOGY?

One question that might be asked is 'Why do we need another new methodology'? Indeed, there are many existing methodologies for designing software. In particular, object-oriented analysis and design have been extensively studied and developed. Is it not possible to use object-oriented techniques to build agent systems?

The short answer is 'Not well!'. Although agents and objects do have similarities, the differences are significant. It is possible to use object-oriented analysis and design techniques to design agent systems. However, the fit is not natural and the resulting design is less likely to make good use of agents.

For example, one important aspect of agents is that they are *proactive*, that is, that they pursue their own agenda over time. This is realized in terms of goals. A methodology that supports proactive agents needs to support the explicit modelling of goals, which is not generally a part of object-oriented methodologies. By contrast, Prometheus models goals and thus supports the design of proactive agents.

Some other areas where Prometheus differs significantly from object-oriented methodologies include the following:

- The provision of a process for determining the types[1] of agents in the system.

- Treating messages as components in their own right, not just as labels on arcs. This allows a message (or an event) to be handled by multiple plans, which is crucial to achieving flexibility and robustness.

- Distinguishing percepts and actions from messages, and looking explicitly at percept processing.
  Agents are situated in an environment, and it is important to define the interface between agents and their environment. Percept processing is often important for agents that are situated in the real world and take their percepts from noisy devices such as video cameras.

---

[1] Agents existing at run time are instances of agent types. For example, in designing an online book store, we may identify agent types such as a Sales Assistant agent type. At run time, there may be multiple agents that are instances of this type.

- Distinguishing passive components (data, beliefs) from active components (agents, capabilities, plans): with object-oriented modelling, everything is modelled as (passive) objects.

- One view of agents (the intentional stance (Dennett 1987)) ascribes *mental attitudes*, such as beliefs, and desires to agents. If we subscribe to this view, then we would like the design methodology to address these aspects. Existing non-agent methodologies do not ascribe mental attitudes to software components. It is worth pointing out that some agent-oriented methodologies (e.g. MaSE) do not subscribe to this view, and, consequently, do not address mental attitudes. Others, including Prometheus, do capture mental attitudes during the analysis and design processes.

Note that although there are clear differences between Prometheus and object-oriented methodologies, there are also commonalities. Although current object-oriented methodologies are not sufficient for engineering agent-oriented software, they *are* relevant – agents are software, and, indeed, many aspects of the Prometheus methodology have been based on object-oriented methods and notations. For example, use case scenarios are adapted from standard practice (Jacobson *et al.* 1992); interaction diagrams are UML sequence diagrams; AUML (`http://www.auml.org/`) (itself an extension of UML) is used directly, and the Rational Unified Process (RUP) (Kruchten 1998) and Prometheus share a similar approach to applying an iterative process over clearly delineated phases.

We have argued that the current mainstream methodologies, as exemplified by UML and RUP, do not provide sufficient support for producing good agent-oriented designs. We now give a brief overview of the agent-oriented methodology Prometheus, which will be the focus of much of this book.

3.2 PROMETHEUS: A BRIEF OVERVIEW

The Prometheus[2] methodology defines a detailed process for specifying, designing, implementing and testing/debugging[3] agent-oriented software systems. In addition to detailed processes (and many practical tips), it defines a range of artifacts that are produced along the way. Some of these artifacts are kept, and some are only used as 'stepping stones'. Some of the artifacts are graphical while others are structured text (i.e. forms).

Prometheus' artifacts relate back to the agent concepts that were introduced in the previous chapter. For example, actions and percepts are captured in the system specification phase; the detailed design phase results in plans, events and beliefs; and the entities used in the various overview diagrams correspond directly to the concepts.

Note that all of the artifacts are structured. This is important in order to be able to provide tool support for the methodology.

The *Prometheus* methodology consists of three phases, depicted in Figure 3.1.

---

[2]Prometheus was the wisest Titan. His name means 'forethought' and he was able to foretell the future. Prometheus is known as the protector and benefactor of man. He gave mankind a number of gifts including fire. (from *http://www.greekmythology.com/*)

[3]Testing and debugging are not covered in this book. Poutakidis *et al.* (2002, 2003) have proposed a debugging method and tool to debug agent interactions, but this is not yet mature enough to be usable for debugging real systems.

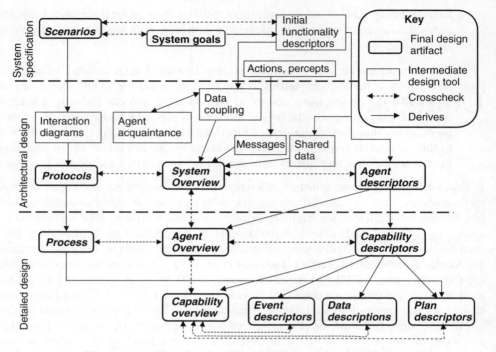

Figure 3.1   The phases of the Prometheus methodology

1. The *system specification phase* focuses on identifying the goals and basic functionalities of the system, along with inputs (percepts) and outputs (actions).

2. The *architectural design phase* uses the outputs from the previous phase to determine which agent types the system will contain and how they will interact.

3. The *detailed design phase* looks at the internals of each agent and how it will
   accomplish its tasks within the overall system.

A fourth phase is implementation, which is omitted from Figure 3.1 because its details
depend on the implementation platform chosen.

The following description of these phases is intended to give a rough feel for the
overall structure of the methodology, so that when reading the following chapters, where
Prometheus is described in detail, you have some idea of how the details fit into the
bigger picture.

3.2.1   SYSTEM SPECIFICATION

The system specification phase, described in detail in Chapter 4, focuses on the following:

- Identifying the **system goals**.

- Developing use case **scenarios** illustrating the system's operation.

- Identifying the basic **functionalities** of the system.

- Specifying the interface between the system and its environment in terms of **actions** and **percepts**.

These four steps do not proceed in sequence; rather, one shifts between them. For example, after adding a goal, we might add a use case scenario illustrating how this goal is achieved. The scenario might include new goals and/or functionalities.

System goals are useful in capturing, at a high level, what the system needs to be able to do. Because goals are a fairly high-level description, they tend to be less likely to change over time than functionalities. Identifying the goals results in a collection of goals, each with a name and description. We also capture the relationship between goals and sub-goals.

Use case scenarios are examples of the system's operation. They are useful in that they are more specific and concrete than goals and thus tend to be more accessible and easier to visualize. However, they do not tell the whole story. Roughly speaking, a use case scenario consists of a sequence of steps that occurs during the system's operation, along with a description of the context in which this sequence of steps could occur. Use case scenarios are borrowed from object-oriented design; however the details differ.

Functionalities are chunks of behaviour formed by grouping related goals, together with related data, percepts and actions. For example, in defining an electronic bookstore we may include functionalities such as "Delivery Handling Functionality" or "Stock Purchasing Functionality". These functionalities should be specific enough that they can be described adequately in a sentence or two.

We also specify the environment in which the agents will be situated. This is defined in terms of *actions* (ways that the agents affect the environment), and *percepts* (incoming information from the environment).

3.2.2 ARCHITECTURAL DESIGN

The architectural design phase, detailed in Chapters 5, 6 and 7, focuses on the following:

- Deciding what **agent types** will be implemented and developing the **agent descriptors**.

- Capturing the system's overall (static) structure using the **system overview diagram**.

- Describing the dynamic behaviour of the system using **interaction diagrams** and **interaction protocols**.

A major decision to be made during the architectural design is which agent types should exist. This is done by grouping functionalities into agent types. Each agent type consists of one or more functionalities. Given a set of functionalities, there is a large number of possible groupings. Deciding on a reasonable grouping is guided by considerations of coupling and cohesion. A **data coupling diagram** is used to help guide this decision, and an **agent acquaintance diagram** is used to help assess whether a reasonable decision was made. Once a grouping is chosen, each agent type is described using an **agent descriptor** form.

Once we have decided upon the agents in the system, we identify which agents react to which percepts, as well as which agents perform particular actions on the external environment. In addition, we specify the **messages** between agents and determine the major **shared data** repositories. These items form the overall design of the system and are depicted in the **system overview diagram**. The system overview diagram is perhaps the single most important product of the design process. It ties together agents, data, external input and output, and shows the communication between agents.

The system overview diagram shows the pathways of communication – which agents talk to which other agents – but not the *timing* of communication – which messages are followed by which other messages. The timing of communication is captured initially in the **scenarios**: whenever an activity by an agent is followed by an activity of another agent, there is, implicitly, communication between them. This communication is depicted explicitly in **agent interaction diagrams**. Like scenarios, interaction diagrams depict one possible sequence of messages between agents. In order to describe all possible interactions, we develop **interaction protocols**, depicted using Agent UML (AUML[4]) (Huget *et al.* 2003).

3.2.3 DETAILED DESIGN

Detailed design focuses on developing the internal structure of each of the agents and how it will achieve its tasks within the system.

In essence, we progressively refine each agent by defining **capabilities** (modules within the agent), internal **events**, **plans** and detailed **data** structures. This process begins by describing agents' internals in terms of capabilities. The internal structure of each capability is then described, optionally using or introducing further capabilities[5]. These are refined in turn until all capabilities have been defined. At the bottom level, capabilities are defined in terms of plans, events and data. In parallel, we continue to work on developing the dynamics of the system by refining interaction protocols into **process specifications**.

The detailed design process presented in this book is split into two parts. In Chapter 8, we focus on the following:

- The refinement of agents in terms of capabilities, giving the **agent overview diagram** and **capability descriptors**; and

- The development of **process specifications**.

This part of the process neither assumes nor relies on any particular implementation platform or architecture.

---

[4]We use the revised version of AUML currently being developed. For a brief introduction, see Appendix C.

[5]Capabilities are allowed to be nested within other capabilities and thus this model allows for as many layers within the detailed design as are needed in order to achieve an understandable complexity at each level.

The agent overview and the capability descriptors are analogous to the system overview diagram and the agent descriptors, but now focussing within a single agent. The process specifications provide a detailed view of an individual agent's part in a particular process, as defined by the protocol specification at the global level. Each global protocol will have a number of corresponding local views, each of which defines the process from the point of view of a particular agent.

In Chapter 9, we continue the detailed design process, focussing on:

- Design of the plans within a capability and the events generated and handled by these plans, as captured in the **capability overview diagrams**.

- Specification of the algorithm within each plan, as well as associated data (or beliefs) and detailed specification of events. These are captured in **plan, data and event descriptors**.

The capability overview diagrams are similar in style to the agent overview and system overview diagrams, although plans must indicate which incoming event is the *trigger* event. Plans are part of the specification of the dynamics of the system and in developing them we take into account the process specifications, as well as structural information such as what triggers them. The descriptors provide the details necessary to move into implementation. Exactly what are the appropriate details for these descriptors will depend on aspects of the implementation platform.

It is only in this final part of the detailed design that we are committing to a particular implementation style. Specifically, we are assuming that plans are triggered by events and that it is possible to have multiple plans that handle a given event type, where the choice of plan to be used is determined at run time. This assumption corresponds to a whole class of implementation platforms (see Chapter 10) including BDI systems such as JACK and systems based on hierarchical task networks such as RETSINA.

3.3 GUIDELINES FOR USING PROMETHEUS

Before we discuss the details of Prometheus, we discuss a few issues involved in using Prometheus:

- Not following the methodology strictly

- The role of iteration and tool support

- When to use agents.

The Prometheus methodology is intended to be interpreted as a set of *guidelines*, not followed strictly. In developing Prometheus, we have had to steer a fine balance between committing to certain ways of doing things and keeping the methodology from becoming too complex and too specific. We hope that Prometheus, as is, will be useful for a large range of domains and users. However, we do not expect that Prometheus will be a perfect fit for all situations. For example, a small and simple agent system may not need to use capabilities; or an agent system may be simulating an existing (human) organization and thus the agent types may already be known, making the use of functionalities and of the data coupling diagram redundant.

In addition, Prometheus is still evolving. There are a number of types of systems that it does not yet handle well. For example, Prometheus as described in this book does not address agent teamwork or mobile agents.

The basic principle is that you should use your common sense and judgement. If you have less experience in the design of agent systems, you will probably want to follow all of the process steps described. However, as you become more experienced, you should use your judgement in applying the methodology.

ITERATIVE DEVELOPMENT

A key element of modern software engineering is *iteration*. Like any other complex human endeavour, such as writing a book, it is not possible to get everything right the first time. Instead of attempting to get things perfect before moving on (as in the waterfall model), modern software engineering processes such as RUP are iterative: although there are still clearly defined activities such as requirements specification, high-level design, detailed design, implementation and testing, these activities are not done in sequence a single time. Rather, the whole process is iterated and the emphasis gradually shifts. The first few iterations might involve primarily requirements-specification activities; however, later iterations will introduce other activities. Prometheus also adopts this approach. Although the following chapters describe the phases and activities in a sequential manner, we do not advocate that they be applied sequentially.

One issue that arises with any iterative process that modifies an existing design is that changes in one part of a design can introduce inconsistencies. This necessitates cross-checking as the design is developed and modified. Although it is possible to perform such cross-checking manually, this is tedious and, more importantly, error prone. Fortunately, it is possible to provide automated support that can detect certain inconsistencies. Our experience has been that this sort of automated cross-checking is very useful in an iterative process.

WHEN TO USE AGENTS

It is important to consider what parts of a system should be treated as agents and designed using an agent-oriented methodology (such as Prometheus), and also how the links between an agent-oriented sub-system and non-agent software can be designed and implemented.

Not all software components are best viewed, modelled and designed as agents. Sometimes, you will be designing a system where it makes sense to model it entirely as a multi-agent system. However, this is not always the case. Some sub-systems may not make sense or may not benefit from being viewed as a collection of agents. For example, an image-processing sub-system that extracts the position of a ball from video frames will not benefit from being viewed as an agent or as a system of agents.

How can we identify which parts of a system should be viewed as agents (and which parts should not)? The short, pragmatic answer is that we should use agents where they are more natural and offer a benefit. The following questions can be used to help identify components that should be treated as agents (if the answers are 'yes'):

- Is it autonomous?

- Does it have goals?

- Viewed as an object, is it active (in the sense of having internal threads that run concurrently with the rest of the system)?

- Does it do multiple things at once? If so, does it need to reason about interaction between the different activities?

- Does it need to change the way it is doing things on the basis of changes in its environment?

If the answers to these questions are mostly 'yes', then you should probably think of the components as agents (and design them accordingly).

EXTERNAL DATA AND CODE

Assuming that some part of the system will be designed as a multi-agent system using Prometheus and some other parts will be designed using other methods, how can the designs be integrated? There are two approaches: we can either view a non-agent part as being an external sub-system with which the agent system as a whole interacts, or as being an ability of a particular agent type.

The system overview diagram used in Prometheus allows for external data and for external 'library code' to be specified. External data can be accessed by various agent types, and this is specified in the design of the agent system. External library code can also be used by agents. The precise mechanism varies: it can be viewed as communicating with the agent system via messages, or as delivering percepts and accepting action requests. The specification of the external sub-system, if it is being built (as opposed to simply using existing code), can be done using whatever methodology is appropriate: object-oriented design, database design, or some other established method.

Alternatively, we can view non-agent parts as being *within* a certain agent type. Agents can include data. This data can be specified, structured and designed in many ways. Some data might be best viewed as a relational table, other data might be best viewed as objects. In very simple cases, data might just be provided data types (e.g. an agent's velocity could just be a floating point number). In addition to including data, agents can also include any existing code by including the appropriate function call within an agent plan. This is viewed as an "internal action".

3.4 AGENT-ORIENTED METHODOLOGIES

A large number of agent-oriented methodologies have been proposed in recent years (Brazier *et al.* 1997; Bresciani *et al.* 2002; Burmeister 1996; Burrafato and Cossentino 2002; Bush *et al.* 2001; Caire *et al.* 2001; Collinot *et al.* 1996; Cossentino and Potts 2002; Debenham and Henderson-Sellers 2002; DeLoach *et al.* 2001; Drogoul and Zucker 1998; Elammari and Lalonde 1999; Giunchiglia *et al.* 2002; Glaser 1996; Iglesias *et al.*

1999, 1997; Kendall *et al.* 1995; Kinny and Georgeff 1996; Kinny *et al.* 1996; Lind 2000; Odell *et al.* 2000; Shehory and Sturm 2001; Varga *et al.* 1994; Wagner 2002, 2003; Wooldridge *et al.* 2000). The aim of this book is to describe Prometheus, not to survey the many existing methodologies. Thus, this section is intentionally brief and incomplete. We encourage the interested reader to read the publications describing the various methodologies and the comparisons between methodologies that are beginning to appear (Cernuzzi and Rossi 2002; Dam 2003; Dam and Winikoff 2003; O'Malley and DeLoach 2001; Shehory and Sturm 2001; Sturm and Shehory 2002, 2003).

PROMETHEUS DESIGN GOALS

Prometheus is intended to be useful and usable by industry developers. In order to be usable, it must be described in sufficient detail and be complete, that is, cover all the necessary activities and phases. In order to be useful, it must support the design of realistically sized systems and thus must support an iterative mode of application (and, in particular, cannot assume a 'waterfall' process model). A consequence of this is that tool support is, in our opinion, highly desirable. We also believe that in order to be useful, Prometheus must focus on designing agents that are flexible and robust. In particular, the use of goals and plans as an implementation technology allows agents to be flexible and robust.

Thus, the design of Prometheus was guided by the following design criteria, aimed at ensuring that the methodology be useful and usable:

- Prometheus must be detailed enough to use and must be complete.

- Prometheus must support the development of agents based on goals, plans and beliefs.

- Prometheus must scale to large systems, and must not use a waterfall process model.

- Prometheus must facilitate tool support, and such tool support should be implemented.

Additionally, in order to ensure that Prometheus is actually useful to, and usable by, industry developers, it must be used by industry developers! In our work, we have both worked closely with the company Agent Oriented Software and have also taught Prometheus to undergraduate students.

These criteria set Prometheus apart from existing methodologies. Many of the existing agent-oriented methodologies are not yet ready to be used by industry developers: they are still under research, and either focus on specific aspects of agent design such as teamwork (i.e. do not provide a complete methodology), or are not described in sufficient detail. For example, many of the methodologies are only described in a small number of short (8 to 15 page) conference papers. Additionally, many methodologies do not provide tool support.

SOME SPECIFIC METHODOLOGIES

Of the long list of agent-oriented methodologies, there are some that *are* described in detail, that do offer tool support and that do appear to be ready for use. In particular, the

MaSE and Tropos methodologies are both complete, have been developed over a period of time (i.e. are mature) and both provide detailed descriptions.

The **Gaia** methodology (Wooldridge *et al*. 2000), like Prometheus, has been developed over a number of years by people experienced in building agent systems. However, we have found that the lack of a detailed design process – intentionally absent because of a desire for generality – meant that it did not provide sufficient support for the needs of those we were working with. There are similarities between Prometheus and Gaia for specification and architectural design. Our agent acquaintance diagrams are essentially the same as those used by Gaia, and the roles of Gaia are similar in concept to functionalities in Prometheus, although there are slightly different things that are considered.

The **Tropos** methodology (Bresciani *et al*. 2002; Giunchiglia *et al*. 2002) covers early requirements to detailed design. Its detailed design is oriented very specifically towards JACK as an implementation platform. Compared to Prometheus, Tropos provides an early requirements phase, which Prometheus does not (although it would certainly be possible to adapt Tropos' early requirements phase for use in Prometheus). Prometheus provides a more detailed process – particularly in the architectural design phase. Prometheus also provides tool support and cross-checking; tool support for Tropos is currently only in the form of a diagram editor rather than the consistency checking and automatic generation of some parts of the design that is part of the Prometheus Design Tool.

The **MaSE** methodology (DeLoach *et al*. 2001) is one of the few methodologies that appears to have significant tool support. However, MaSE is unsuitable for our purposes since it views agents ' ... merely as a convenient abstraction, which may or may not possess intelligence' (DeLoach *et al*. 2001, p. 232). Thus, MaSE (intentionally) does not support the construction of plan-based agents that are able to provide a flexible mix of reactive and proactive behaviour. Rather, MaSE aims to be general and treats agents as 'simple software processes that interact with each other to meet an overall system goal' (DeLoach 2001, p. 232).

PASSI (Burrafato and Cossentino 2002; Cossentino and Potts 2002) is a recent addition to the list of methodologies. Although it has not, to the best of our knowledge, been described in detail, it appears to be complete, and provides tool support.

COMPARISONS

Comparisons of the existing methodologies are limited, but are beginning to appear. MaSE, Prometheus and Tropos are compared using a feature-based approach in which the assessment of each methodology against the criteria is validated using a survey of the developers of the methodology (and of students) (Dam and Winikoff 2003). This work is extended to include MESSAGE and GAIA and to also provide a comparative analysis of the models and processes of each of the methodologies (Dam 2003). Other comparisons between agent-oriented methodologies include (Shehory and Sturm 2001), (Cernuzzi and Rossi 2002), (Sturm and Shehory 2003).

4

System Specification

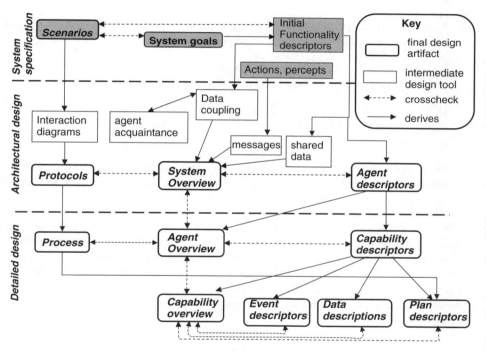

Phases, artifacts and relationships in the design process

This chapter discusses in detail the artifacts and processes in the System Specification phase, the initial phase of the Prometheus methodology. Often, the initial documents from which a system analyst or system developer begins are simply a few paragraphs of loose description, or a rough understanding based on discussions and meetings. From this, a

Developing Intelligent Agent Systems L. Padgham & M. Winikoff
© 2004 John Wiley & Sons, Ltd ISBN: 0-470-86120-7 (HB)

clear and precise understanding of the system to be built must be developed. Preferably, it is developed in a manner that facilitates clear and traceable paths between the system specification phase and subsequent system development phases, such as design, implementation and testing.

Considerable work has been done in recent years on requirements engineering, recognizing that care and rigour at this early stage can avoid many more expensive problems at later stages. The Prometheus methodology focuses particularly on specification of goals, using some of the work of van Lamsweerde (e.g. (van Lamsweerde 2001)), and on scenario descriptions. In addition, it requires specification of *functionalities* (small chunks of behaviour) related to the identified goals. There is also a focus on the important issue of how the agent system interfaces with the environment in which it is situated, in terms of *percepts* that arrive, or can be obtained, from the environment, and *actions* that impact on the environment. As part of the interface specification, we also address interaction with any external data stores or information repositories.

The aspects that we develop in the system specification phase are then as follows:

1. Specification of system goals, resulting in a list of goals and sub-goals, with associated descriptors.

2. Development of a set of scenarios that have adequate coverage of the goals, and which provide a process-oriented view of the system to be developed.

3. Definition of a set of functionalities that are linked to one or more goals, and capture a limited piece of system behaviour, which can be described in a few sentences.

4. Description of the interface between the agent system and the environment in which it is situated, in terms of incoming percepts, outgoing actions and external information stores with which the system will interact.

These activities interact to support and complement each other, and the developer typically moves between them in a flexible manner. The productive interaction between scenario descriptions and goal elicitation has been explored increasingly in the requirements engineering literature in recent years (Liu and Yu 2001; Rolland *et al.* 1999). Our experience in agent system design is also that this interaction, together with specification of system interfaces, is natural and productive.

4.1 GOAL SPECIFICATION

The reasons for building the system should always be central in one's thinking when specifying the system, and thus *goals* are a natural construct to use in system specification. In addition, goals are central to the functioning of the intelligent software agents that are going to realize the system. The use of goals at the requirements engineering or system specification phase thus facilitates a mapping into later detailed design and implementation, as well as providing an appropriate mechanism for requirements specification.

4.1.1 IDENTIFY INITIAL GOALS

The initial brief system description usually contains some implicit indications of the system goals – what it is that the system is supposed to do. These provide a starting point for building an initial list of system goals. A simple underlining or highlighting of the relevant words or phrases in the description can be used to provide a first cut at the system goals.

For example, consider the following description:

> We wish to develop a student enrolment system that allows students to enrol in subjects, add and delete subjects in accordance with rules and view their enrolment. Enrolment rules should be editable by authorized staff.

On the basis of this initial description, it is possible to extract the following system goals:

- Enrol student

- Add subject to enrolment

- Delete subject from enrolment

- Check enrolment against rules

- View enrolment

- Authenticate staff

- Edit rules.

These can then provide a basis for refinement, which provides an extended list of goals that can then be grouped into what we call *functionalities* – small chunks of system behaviour, describable in a few sentences.

In the rest of this chapter and the following chapters, we will develop an example of an electronic bookstore to illustrate the design process. To enable easy following of the example, we will enclose all of the example details in a framed box (which may extend over page breaks, in which case the bottom and top of the frame on the adjacent pages will be missing), as below. In addition, the collected details of the example can be found in Appendix A.

Electronic Bookstore: Case study

Here, we show an initial brief system description of an electronic bookstore and the set of system goals that can be extracted from that description:

> We would like to develop a fully online system for worldwide sale of books. This system will offer a broad range of books to customers, and a personalized, friendly user interface. The system must facilitate fast and reliable service at all stages, from locating a desired book, to delivery of the purchase. The store should have competitive prices.

We identify our initial set of goals by highlighting parts of this description:

We would like to develop a **fully online system** for **worldwide sale of books**. This system will offer a **broad range of books** to customers, and a **personalized, friendly user interface**. The system must facilitate **fast and reliable service** at all stages, from **locating** a desired book, to **delivery** of the **purchase**. The store should have **competitive prices**.

This then yields the following extracted system goals:

- Worldwide sale of books

- Fully online system

- Broad range of books

- Personalized, friendly user interface

- Fast and reliable service

- Locating of books

- Delivery of books

- Purchase of books

- Competitive prices.

4.1.2 GOAL REFINEMENT

We now refine these goals somewhat using the simple technique of asking 'how?' as suggested by (van Lamsweerde 2001). We consider each goal and ask 'how might this goal be achieved?'; the answers give the *sub-goals* of the goal under consideration. For example, the goal of providing worldwide sale of books could be achieved by having an online system and by delivering books internationally.

It would be possible to also use the abstraction techniques of asking 'why?', espoused by the same author, to build up a much more complete goal tree. However, we have chosen to limit the set of goals to those arising most directly.

As we refine the original goals to be slightly more specific, it is often the case that we find similar sub-goals arising under different initial goals. A grouping of similar sub-goals then provides the basis for what we call 'functionalities' – descriptions of limited chunks of system behaviour.

For example, refining the original goal of *fully online system*, we obtain sub-goals *find books online*, *pay online* and *order online*, while the original goal of *purchase books* leads to *find books*, *place order*, *make payment* and *arrange delivery*. *Pay online* and *make payment* are clearly closely related, if not identical, goals, and are therefore grouped together. Below we show the expanded goal set derived from our initial set of goals, and the initial arrangement of these into groupings.

Electronic Bookstore: Case study

Here, we have the expanded list of goals and associated sub-goals resulting from asking 'how?' and refining the initial list.

- Worldwide sale of books.
  - online system
  - deliver internationally

- Fully online system.
  - find books online
  - order online
  - pay online

- Broad range of books.
  - books from many publishers
  - update catalogue regularly
  - multiple suppliers

- Personalized, friendly user interface.
  - personalized welcoming
  - recommendations based on user profile
  - information available about orders in process

- Fast and reliable service.
  - arrange courier deliveries
  - provide estimates of delivery time
  - track delivery problems
  - have books available in stock

- Locating of books.
  - provide search facility
  - provide recommendations

- Delivery of books.
  - arrange delivery
  - monitor delivery

- Purchase books
  - find books
  - place order
  - make payment
  - arrange delivery

- Competitive prices.
  - set prices competitively
  - temporarily reduce prices to match competitors

After refining the goals, we can then start to rearrange them, moving similar goals together. As we do this we may find the same sub-goal under different parent goals, in which case we can coalesce them. We can then give a name to the grouping which describes it. This may be related to one of the original top level goals, or may be a new name.

Electronic Bookstore: Case study

Below is an initial re-arrangement of the goals with moved sub-goals shown in *italics*. The name chosen for the resulting grouping is shown in CAPITALS below each group of sub-goals. In some cases such as *competitive prices*, the sub-goals are divided into two functionalities.

- Worldwide sale of books

- Fully online system.
  - *online system* (from Worldwide sale of books)
  ONLINE INTERACTION

- Broad range of books.
  - books from many publishers
  - update catalogue regularly
  CATALOGUE MANAGEMENT

- Personalized, friendly user interface.
  - personalized welcoming
  WELCOMING
  - recommendations based on user profile
  - *provide recommendations* (from Locating of books)
  PROFILE MONITOR

- Fast and reliable service.
  - have books available in stock
  - *multiple suppliers* (from Broad range of books)
  STOCK MANAGEMENT

- Locating of books.
  - *find books online* (from Fully online system and also *find books* from Purchase books)
  - provide search facility
  BOOK FINDING

- Delivery of books.
  - *deliver internationally* (from Worldwide sale of books)
  - *arrange courier deliveries* (from Fast and reliable service)
  - arrange delivery (also includes *arrange delivery* from Purchase books)
  - monitor delivery
  - *provide estimates of delivery time* (from Fast and reliable service)

- *track delivery problems* (from Fast and reliable service)
- *information available about orders in process* (from Personalized, friendly user interface)
DELIVERY HANDLING

- Purchase books
  - place order
  - *order online* (from Fully online system)
  - make payment
  - *pay online* (from Fully online system)
  PURCHASING

- Competitive prices.
  - set prices competitively
  PRICE SETTING
  - temporarily reduce prices to match competitors
  COMPETITION MANAGEMENT

We continue to work with the list of goals and sub-goals, coalescing similar goals and adding goals as we see that a particular grouping is lacking some aspect. For example, the sub-goals of *deliver internationally*, *arrange courier delivery* and *arrange delivery* are coalesced into the single sub-goal *arrange delivery*.

As we do this, our original set of goals and sub-goals becomes a network of connected goals as shown in Figure 4.1. Each goal is represented by an oval and arrows join goals to sub-goals.

We also elaborate goals by adding more detailed sub-goals; for example, we add to PROFILE MONITOR the sub-goals of registering and updating the customer profile. In some cases, a sub-goal is replaced with a more detailed sub-goal (or collection of sub-goals). For example, under STOCK MANAGEMENT, we replace *have books available in stock* with *reorder stock*, and sub-goals to log outgoing and arriving books.

We aim for groupings of approximately two to five goals that belong together in a way that we can provide a brief but comprehensive description of the functionality to which these goals belong.

Some of our original goals may disappear if all sub-goals belonging to them are moved into other groupings. In our example, this happens for the original goal of *worldwide sale of books* as the sub-goals indicating how this goal was to be realized – *online system* and *worldwide delivery* – are moved. Similarly, some original goals may expand into more than one grouping, as is seen with *Delivery of books*, which leads to two groupings – one dealing with arranging deliveries and the other dealing with delivery problems. We may choose to retain goals such as *worldwide sale of books*, but mark them as *abstract*, indicating that they are a motivator for more concrete goals, but will not appear directly as the details are developed.

Below is the revised set of goal groupings for our example, organized by headings.

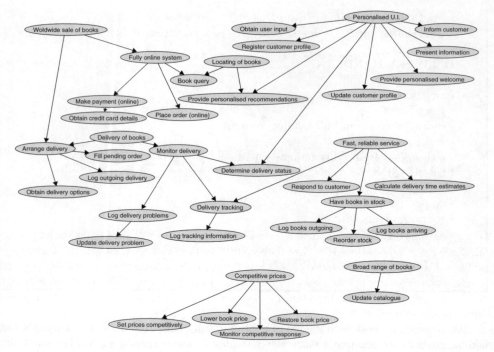

Figure 4.1   Goals for the electronic bookstore

Electronic Bookstore: Case study

The set of goal groupings that we developed after our initial pass at refining, coalescing and adding goals as needed:

- ONLINE INTERACTIONS
  - obtain user input
  - present information

- CATALOGUE MANAGEMENT
  - update catalogue
  (note that a decision has been made at this point that the goal of having books from many publishers will be met outside the software system. The system will simply manage the catalogues it is given. Ensuring that these come from a wide variety of sources becomes the responsibility of the company using the system.)

- WELCOMING
  - provide personalized welcome

- PROFILE MONITOR
  - provide personalized recommendations

- register customer profile
- update customer profile

- STOCK MANAGEMENT
  - log books arriving
  - log books outgoing
  - reorder stock

- BOOK FINDING
  - book query

- DELIVERY HANDLING
  - obtain delivery options
  - calculate delivery time estimates
  - arrange delivery
  - determine delivery status
  - log outgoing delivery

- LOST GOODS MANAGEMENT
  - log delivery problem
  - request delivery tracking
  - respond to customer[1]
  - log tracking information

- PURCHASING
  - purchase of books
  - place order
  - make payment

- PRICE SETTING
  - set prices

- COMPETITION MANAGEMENT
  - lower book price
  - monitor competitive response
  - restore book price

4.2 FUNCTIONALITIES

Functionality is the term we use for a chunk of behaviour, which includes a grouping of related goals, as well as percepts, actions and data relevant to the behaviour. A functionality should be coherent, in that it can be described adequately in one or two sentences, and can be named in a way that captures its essence. Functionalities allow for a mixture of both top-down and bottom-up design. They are identified by a top-down

---

[1] . . . with the status of the goods ordered.

process of goal development. At the same time, they provide a bottom-up mechanism for determining the agent types and their responsibilities.

The process of refining and then grouping goals suggests an initial set of functionalities. Further work with scenarios and development of the specification may well suggest additional ones. System specification is an iterative process, starting with identification of goals, but then moving between scenarios, goals and functionalities, as well as identifying actions, percepts and data.

The groupings, namings and additions and merging of goals are continually refined. Figure 4.2 shows the current revised functionalities developed for the Electronic Bookstore at this stage. Functionalities are depicted by rectangles, goals with ovals, and actions with an action icon (a rectangle extended with a triangle pointing to the right). Arrows link functionalities to their goals and actions.

Once we have identified functionalities we can begin to develop *functionality descriptors*. Each functionality should be able to be described in a few sentences. If the description is larger than this, then the functionality should be split into multiple functionalities. In addition to a natural language description and information about the goals and actions that are included, the functionality descriptor should also include what we

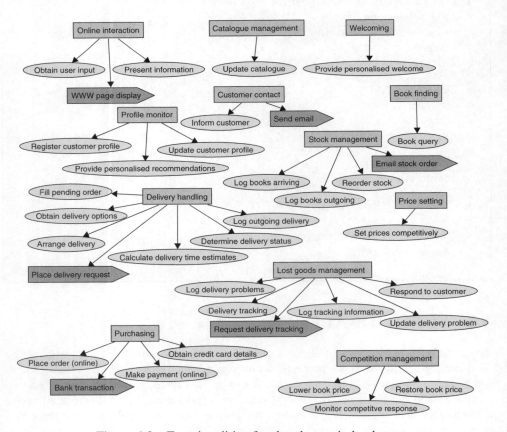

Figure 4.2   Functionalities for the electronic bookstore

call *triggers*: information about what events or situations will cause activity to be initiated within this functionality. These triggers may include identified percepts, but are not limited to these. Notes about the data required and produced by the functionality should also be included.

An example of a functionality descriptor for *Purchasing* is given below.

Electronic Bookstore: Case study

This is an example functionality descriptor that was developed during system specification, on the basis of the three goals grouped under *Stock Management*.

Stock Management Functionality

Description: This functionality monitors the stock available as it comes in and goes out, ordering new stock as needed. It maintains information as to when stock is expected to arrive.

Goals: Reorder stock, Log books arriving, Log books outgoing

Actions: E-mail stock order

Triggers: Stock arrival, Stock order delay, Failed stock arrival

Information used: Stock database, Customer order, Stock order

Information produced: Stock database, Delayed orders, Arrived orders

---

4.3 SCENARIO DEVELOPMENT

Scenarios are complementary to goals in that they show the sequences of steps that take place within the system. In developing goals, we are typically already building up scenarios of how these goals will be part of various processes within the system. Scenarios enable us to specify some of this structure, which in turn may help identify missing goals.

Scenarios are used primarily to illustrate the normal running of the system, although it can also be useful to develop some scenarios that indicate what is expected to happen when something goes wrong. As scenarios are developed, it becomes evident where there is a need for information from the environment (i.e. percepts) and where actions are required. Also, as scenarios are developed, it is common to identify additional goals that are needed. For example, in developing the **Query Late Books Scenario** in which the user is querying because their books have not arrived, we identified the need for a new goal *Update delivery problem*. We already had *Log delivery problem*, but discovered the need to update the information once the information from the tracking request had been received. Also in this scenario, we can see that two percepts and an action are required. The two percepts are the original user query from the web interface and the response to the tracking initiated, which are both input to the system in some way. The action is the request for tracking of the delivery, which goes outside the system and initiates behaviour in the environment. Below we show the outline of this scenario made up of existing goals, the newly identified goal, the percepts and the action.

The core of a scenario consists of a sequence of steps. Possible steps are achieving a goal (GOAL), performing an action (ACTION), receiving a percept (PERCEPT), or referring to another use case scenario (SCENARIO). Additionally, we use the step type OTHER to cover unusual steps such as waiting for something to happen.

Electronic Bookstore: Case study

Here is the outline of the **Query Late Books Scenario** developed in the first round of scenario descriptions:

Query Late Books Scenario
Trigger: user enquiry
Description: When the user enquires about the status of their delivery the system requests that tracking be performed and waits for the result. In this scenario the result is that the book could not located and so a replacement book is sent.

1. GOAL: Determine delivery status

2. GOAL: Log delivery problems

3. ACTION: Request delivery tracking

4. GOAL: Inform customer

5. OTHER: wait for response ...

6. PERCEPT: tracking info received

7. GOAL: Arrange delivery

8. GOAL: Log books outgoing

9. GOAL: Inform customer

10. GOAL: Update delivery problem

In the process of developing scenarios, it is common to find the need for new GOAL steps. These are then added to the appropriate functionality, if there is a natural place where they belong. It may be the case that there is a need to also introduce a new functionality. This can happen either because a new goal does not naturally belong within any of the existing groupings or if a grouping becomes too large, so there is a need to divide it. For example, in the process of developing the above scenario, the new goal *Inform Customer* was identified. This was considered for inclusion with the goals in the *Online Interaction* functionality. However, since informing the customer could be done either via the web or by email, and since email is not an *Online Interaction*, it was decided to have two different functionalities for interacting with the customer, and a *Customer Contact* functionality was introduced.

4.3.1 GOAL STEP DETAILS

We also attach to each GOAL step, the name of the functionality it belongs to, the information that is used, and the information that is produced or written. So for the step *Arrange delivery*, we also attach the name of the functionality (Delivery Handling), the information used (Order record) and the information produced (none). For example[2],

Num.	Step type	Step	Functionality	Data used and produced
5.	GOAL	*Arrange delivery*	Delivery Handling	Customer Order Customer Order

Below we show the steps of the **Query Late Books scenario**, with the full information for each step.

Electronic Bookstore: Case study

Fully developed steps for the *Query Late Books* scenario:

Key for functionality and data abbreviations:

DH Delivery Handling
LGM Lost Goods Management
CC Customer Contact
SM Stock Management
Cust. Info.: Customer Information
Cust. Order: Customer Order
Del. Problem: Delivery Problem

	Step type	Step	Functionality	Data used and produced
1	GOAL:	*Determine delivery status*	DH	Cust. Order none
2	GOAL:	*Log delivery problems*	LGM	Cust. Order Del. Problem
3	ACTION:	*Request delivery tracking*	LGM	Cust. Order Del. Problem

---

[2]In order to avoid tables that extend past the width of the page, we put the data used on the first line and the data produced on a second line.

4	GOAL:	*Inform customer*	CC	Del. Problem
				Cust. Info.
5	OTHER:	Await response		
6	PERCEPT:	*tracking info received*		
7	GOAL:	*Arrange delivery*	DH	Cust. Order
				Cust. Order
8	GOAL:	*Log books*	SM	Book Info
		outgoing		Stock DB
9	GOAL:	*Inform customer*	CC	Cust. Order
				Cust. Info.
10	GOAL:	*Update delivery*	LGM	Track. Resp.
		problem		Cust. Order (used)
				Del. Problem

As can be seen, the above scenario is in two pieces. After an action requesting tracking of the delivery, nothing more can be done until a response is received (or a sufficient delay has occurred to conclude that no response will be received). The scenario could be split into two pieces, with the second piece being triggered by the arrival of the tracking information, or a sufficient time delay. We have chosen to keep it as a single scenario in order to more easily see the entire chain of events arising from the initial trigger. We have used the step type 'OTHER' in order to allow us to indicate the delay and to clarify that there is a wait at this point in the scenario.

4.3.2 CAPTURING ALTERNATIVE SCENARIOS

Scenarios describe a single sequence of steps that can occur. We partially capture alternatives using two techniques. The first technique is that with each scenario we include a description of possible alternatives. These are usually minor variations on the scenario that can be easily described.

Electronic Bookstore: Case study

Alternatives for the *Query Late Books* scenario:

Alternative 1: Book is within or close to expected estimate for delivery time so no tracking is requested and steps 6 to 9 are deleted.

Alternative 2: The tracking information received indicates that the book is on its way, so there is no need for a new delivery. Steps 7 and 8 are deleted.

Alternative 3: After a delay (and possible further requests and delays), no response is received to the tracking request. This situation replaces the percept at step 6, but other steps remain the same.

The second technique for capturing alternatives is to use a collection of scenarios that all relate to a single underlying process. For example, one scenario might describe a normal book order (with minor variations) and another might describe a book order in which the book cannot be obtained and the order is refunded.

Scenarios are intended to give a clearer idea of the system, not to fully define it. Given that there is no attempt to describe all possible scenarios, it may be difficult to know when enough have been developed. There should be a scenario written for at least one version of each important process within the system. A good check is to ensure that all goals, actions and percepts are included in at least one scenario. It is often useful to develop one or two scenarios showing the system functioning when something unexpected happens or an error occurs, to give a sense of how these things will be handled. However, it is usually not productive to define large numbers of exceptional scenarios, as these can clutter the design document and actually make it more difficult to understand the planned system.

Electronic Bookstore: Case study

The list of scenarios developed for the electronic bookstores is as follows:

- Book finding scenario
- Order book scenario
- Pending order arrives scenario
- Stock order scenario
- Stock arrival scenario
- Stock delayed scenario
- Missed stock arrival scenario
- Query late books scenario
- Order status query scenario
- Customer profile update scenario
- WWW site arrival scenario
- Cheaper price notification scenario
- New catalogue scenario

4.4 INTERFACE DESCRIPTION

Agent systems are typically situated in a changing and dynamic environment that can be affected, though not totally controlled, by the agent system. An early question that must be answered is, how is the agent system going to interact with the environment? Specifically, what input about the environment will be available to the agent system while it is running; and what will the agent system do to interact with and affect the environment? In line with standard texts on agents (Russell and Norvig 1995), we call the incoming information 'percepts' and the mechanisms for affecting the environment 'actions'.

4.4.1 PERCEPTS AND ACTIONS

Percepts often require some processing in order to extract the information that is of value to the agent system. For example, vision frames from a robot camera are not in themselves what the agent system needs to reason about. They must first be processed both to extract the symbolic data, such as 'ball at 15, 10; robot at 25, 70' and to extract significant information from this symbolic data. If there is a stream of incoming data, the agent system may want to react only when there is an event of some significance, often determined by change from the previous situation. For example, if a soccer-playing robot sees the ball when it did not do so previously, this is significant. Changes from expectations can also indicate significance. For example, if our robot turns towards where it expects to see the ball and does not see it, this is also significant.

In planning and designing the agent system around the percepts, the designer must take into consideration how data is obtained – does it just arrive, or must the system actively seek the data – as well as the exact nature of the data, and to what extent it can be processed to provide information of interest. It is extremely important to investigate and experiment early on with the exact nature of the percepts available to the system. It is not uncommon for a whole system to be designed and implemented, only to find that the quality of the incoming percepts simply does not allow for reliable provision of the information required by the system.

In systems where percepts originate from physical sensing devices of some sort, the data often contains significant amounts of noise, which may require the use of techniques to filter and cleanse the data. This may be done outside the agent system, however, these issues must be considered carefully at design stage and the developer must establish exactly what it is possible to provide as percepts to the system.

Actions may also be complex, requiring significant design and development outside the realm of the reasoning system, possibly including monitoring for failure or continual feedback loops. This is especially true when manipulation of physical effectors is involved. For example, a system may be designed to have an action *move*, with parameters giving distance and speed. This action will require complex feedback loops within the effector system to enable this action to be carried out. Even electronic actions such as sending a message typically rely on additional low-level code, although this is usually provided in communication interfaces and is not something the developer needs to consider.

Typically, an application has some number of obvious percepts and actions that are the initial definition of how the system will interact with the environment. As the specification is developed, the required interface also becomes further defined. It is extremely important to ensure that these interfaces are correct and/or achievable.

In the electronic bookstore, some initial percepts and actions identified were

- WWW page display (action)

- Bank transaction (action)

- Deliver book (action)

- Arrival at WWW site (percept)

- Bank transaction response (percept)

- Stock arrival (percept)

- User input (percept).

As the specification developed, particularly with the development of scenarios, additional necessary actions and percepts were noted, and some were modified. For example, the electronic system was not able to actually *Deliver book*, so this was changed to an action *Place delivery request*, which sent the delivery request to a courier or to the postal room. The full list of percepts and actions is given below.

Electronic Bookstore: Case study

Following are the percepts and actions identified for the electronic bookstore.

Percepts

- Arrival at WWW site

- Bank transaction response

- Cheaper price report

- Failed stock arrival

- Tracking info

- No tracking response

- Stock arrival

- Stock order delay

- New catalogue

- User input

Actions

- Bank transaction

- E-mail stock order

- Place delivery request

- Request delivery tracking

- Send e-mail

- WWW page display.

As the design develops, decisions may be made to pre-process percepts to provide a number of different percepts to the system. For example, a fairly generic percept such as 'user input' might be replaced by a collection of more specific percepts such as 'user book query', 'user book order', and so on. The original incoming percepts should still be documented, and information should be provided as to what processing is required. However, they may no longer be a direct part of the interface to the agent system.

At this stage, we simply develop lists of percepts and actions, but leave the descriptors and the details until Architectural Design. However, notes may be kept and prototyping done, to determine feasibility.

4.4.2 DATA

As scenarios and functionalities are developed, it is also important to note the data that is a part of these. While developing functionalities and scenarios we note both data that is produced and data that is used. An example of data produced by the Stock Management functionality is the *Stock database*, while an example of data used in a scenario is the *Customer order* data used in the first step of the *Query Late Books* scenario by the GOAL step *Determine delivery status*.

At the system specification phase, we need to especially note any data that is external to the agent system, as this forms part of the interface and should be specified at this stage.

An additional part of the system interface that may need to be specified is the interaction with any other software.

Electronic Bookstore: Case study

In the electronic bookstore, we identify two external data stores that we will use: *Courier DB* and *Postal DB*. These contain information about courier companies' areas/rates, and postal rates respectively.

In addition, we identify the following clusters of information, without as yet defining all their details:

- Customer DB – contains information about customers, their profile, their history of visits to the site and orders.

- Customer Orders – contains records of orders that have been (fairly recently) sent.

- Pending Orders – contains records of orders that have been placed but not yet sent.

- Delivery Problems – contains records of queries about items that have not arrived and the investigation of these situations.

- Books DB – contains a comprehensive listing of books, with information on suppliers, prices, and so on. Not all books are necessarily stocked.

- Stock DB – contains records of books that are stocked.

- Stock Orders – contains records of stock orders placed and awaiting delivery.

4.5 CHECKING FOR COMPLETENESS AND CONSISTENCY

One of the advantages of doing design in a structured way is that it becomes possible to specify a range of checks for consistency and completeness. Many of these can actually be automated, as is done for a number of them in the prototype Prometheus Design Tool.

NAMES SHOULD BE CONSISTENT

One of the most basic but nevertheless important checks is that names should be consistent. A goal that is named *credit check* in one place in the design, should not be called *check credit* in another place. While the designer may think the equivalence is obvious, it can be very confusing for the person trying to understand the design – and certainly it does not facilitate straightforward mapping to implementation. Having a design tool greatly simplifies maintaining naming consistency.

EVERY GOAL IN A SCENARIO AND FUNCTIONALITY

At the system specification phase, the main checks that can be done apart from naming consistency are to ensure that the system is adequately specified with respect to the system goals. It is desirable for all goals to be in some way covered in a scenario. There are three ways a goal can be covered:

- It can be specifically mentioned as a step or a trigger.

- All its sub-goals may be mentioned specifically.

- Its parent goal may be mentioned specifically.

All goals that are not labelled as abstract goals should also be contained in some functionality. This is needed to ensure that they are addressed as the design progresses.

SCENARIOS AND FUNCTIONALITIES FOR EVERY ACTION AND PERCEPT

All percepts and actions should also be shown somewhere in a scenario, and should belong to at least one functionality. If they are missing from the scenarios, then it is possible that some significant behaviour has been overlooked. However, it may also be that it is mentioned in a scenario variation, rather than being specifically covered. For example, if the *bank transaction response* percept had been modelled as an *accept response* and a *reject response*, it is quite possible that the *reject response* would occur only in a scenario variation, rather than in a scenario step.

Checking at this phase is really a matter of drawing the attention of the developer to things that may have been overlooked, rather than identifying things that are clearly wrong or inconsistent. Nevertheless, automated support for this can be very helpful.

5

Architectural Design: Specifying the Agent Types

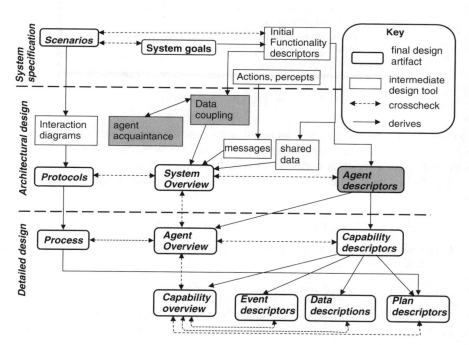

Phases, artifacts and relationships in the design process

In the system specification phase we developed system goals, scenarios and functionality descriptions. In the next three chapters we will cover the *architectural design* phase where we use these artifacts as the basis for developing the high-level design of the agent system. As noted earlier, it is expected that all the phases interact with each other,

Developing Intelligent Agent Systems   L. Padgham & M. Winikoff
© 2004 John Wiley & Sons, Ltd   ISBN: 0-470-86120-7 (HB)

1. Deciding on the agent types used in the application (Chapter 5)

   1.1 Group functionalities into agents considering alternatives

   1.2 Review coupling using agent acquaintance diagrams and decide on a preferred grouping

   1.3 Develop agent descriptors

2. Describe the interaction between agents using interaction diagrams and interaction protocols (Chapter 6)

   2.1 Develop interaction diagrams from scenarios

   2.2 Generalise interaction diagrams to interaction protocols

   2.3 Develop protocol and message descriptors

3. Design the overall system structure (described using a system overview diagram, Chapter 7)

   3.1 Identify the boundaries of the agent system and the interactions with other sub-systems

   3.2 Describe the percepts, and actions, and the relationships between these and relevant agents

   3.3 Define all shared data, both external persistent and internal shared data

   3.4 Develop the system overview diagram

Figure 5.1   The process of architectural design

and that developing the design is an iterative process, also involving implementation. The aspects of architectural design covered in this and the subsequent two chapters interact with each other as well as with system specification.

The three aspects that are developed during architectural design, as shown in Figure 5.1 are:

1. Deciding on the *agent types* used in the application.

2. Describing the interactions between agents using *interaction diagrams* and *interaction protocols*.

3. Designing the overall system structure (described using a *system overview diagram*).

This chapter focuses primarily on the process of deciding on the agent types to be used, while Chapter 6 focuses on defining the interactions, and Chapter 7 covers the system overview.

The architectural design defines what agents are to be a part of the system and how these agents interact with each other to meet the required functionality of the system. It defines the interface of the system to the external environment, and also the interfaces of the individual agents within the system. As the design develops, it is also likely that aspects of the previous stage are revisited and revised, to achieve a cleaner architectural design.

A major decision that is made during the architectural design is the types of agents used. Agent types are formed by combining functionalities. The choice of how functionalities are to be combined is made by considering the functionalities and scenarios and developing possible groupings of functionalities into agents, which are then evaluated according to the standard software engineering criteria of *coupling* and *cohesion*. One proposed grouping of functionalities is better than another if it gives lower coupling and higher cohesion.

> ☞ DEFINITION: *Coupling and Cohesion*
> *Coupling:* Coupling is a property of a group of components. The coupling of components is the extent to which they depend on one another. If a component A depends on component B, then changing B may require changing A as well. Low coupling yields a system in which changes are more likely to be localized. High coupling yields a system in which changing one component is likely to require changes to many other components, thus making the system harder to change. *In Agent systems coupling is exhibited primarily in communication between agents, although use of a shared data store is another possible form of coupling.*
>
> *Cohesion:* Cohesion is a property of a single component. A component is cohesive if all of its parts are related. For example, an object is cohesive if all its data and methods are related to a single clear goal or functionality. *There are a number of ways in which an agent can exhibit cohesion. For example, it could manage a series of actions that occur at the same time (e.g. initialization). An agent could also be responsible for the operations involving a particular data store, including consistency management. Most often, cohesion in an agent is based on the goals of the agent being closely related.*

As with any design, there is a large space of possibilities and no clear right or wrong decisions. What is provided here are some methods and tools to develop and document a design, as well as some ways to evaluate the comparative merits of alternative design options. The *data coupling diagram* and the *agent acquaintance diagram* are tools used to develop and assess proposed groupings.

Once we have decided upon the agents in the system, we identify the pathways of communication (which agents talk to which other agents) as well as the *timing* of communication (which messages are followed by which other messages). The timing of communication is captured initially in the scenarios: whenever an activity by an agent is followed by an activity of another agent, there is, implicitly, communication between them. This communication is depicted explicitly in agent *interaction diagrams*. Like scenarios, interaction diagrams depict one possible sequence of messages between agents. In order to describe all possible interactions, we develop interaction protocols; depicted using the Agent UML (AUML) notation (Huget *et al.* 2003).

Finally, we specify the interactions between the agents, and the system interface in terms of percepts, actions and external data. We also identify any shared data. The overall design of the system is then depicted in the *system overview diagram*, which brings all the items together. The system overview diagram is perhaps the single most important product of the design process. It ties together agents, data, external input and output and shows the communication between agents.

> ☞ TIP: It is possible to show some timing information in a system overview diagram by marking the arcs between agents and messages with numbers. However, our experience has been that, except for very simple cases, this quickly becomes cumbersome and complex. Generally, it is simpler to keep dynamic and static information separate.

5.1 DECIDING ON THE AGENT TYPES

The single most important decision in the architectural design is which agents the system will include. Different options are evaluated using the criteria of *coupling* and *cohesion*. Each agent should be cohesive, in that it can be sensibly described with a short title. The system of agents should be as loosely coupled as possible. It is preferable to have a system in which each agent needs to know about only a limited number of other agents within the system.

The steps we go through in this part of the process are:

1. Group functionalities into agents considering alternative groupings.

2. Review coupling (using agent acquaintance diagrams) and decide on a preferred grouping.

3. Develop agent descriptors.

5.2 GROUPING FUNCTIONALITIES

Given a collection of functionalities, there are obviously many ways in which they can be grouped into agents.[1] At one extreme, every functionality could correspond to a different agent. This is undesirable in that having closely related functionalities in different agents leads to a high degree of dependency, or coupling, between agents. At the other extreme, all of the functionalities could be grouped into a single agent. This is undesirable in that grouping unrelated functionalities together leads to overly complex agents lacking in cohesion.

---

[1] The term 'agent' is somewhat overloaded in that it can refer to both the 'agent type' and to the 'agent instance'. The agent type is a template for agents (similar to a class in object-oriented programming), whereas the agent instance is a run-time entity (similar to an object in object-oriented programming). Context determines which use of agent is intended. At times, we explicitly use 'agent type' or 'agent instance' for extra clarity, but this is cumbersome if used constantly.

The main reasons for deciding to group certain functionalities together into a single agent are:

- The functionalities seem related – it 'makes sense' to group them. For example, the *Price setting and Competition management* functionalities in our case study are clearly related.

- The functionalities require a lot of the same information. If grouped into a single agent, this can then be represented in internal agent data structures. If separate agents are used, the information must be passed via messages unless a shared data store is used, which is not usually a good design decision (see Section 7.4 on page 88).

Reasons for *not* grouping functionalities are primarily the following:

- The functionalities are clearly unrelated

- The functionalities exist on different hardware platforms

- Different numbers of functionalities are required at run time.
  For example, you might have two related functionalities: an *Online Interaction* functionality, and a *Customer Contact* functionality, both of which have to do with customer interaction. However, the *Online Interaction* functionality is required to be part of an agent type where there is one agent per active user. The *Customer Contact* functionality, on the other hand, is required to be in an agent type where there is a single agent of this type in the system. Although the two functionalities are related, we cannot group them together in a single agent type because we cannot simultaneously have exactly one agent of a given agent type and multiple agents of a given type. We term this situation a *cardinality mismatch*.

Other reasons for not grouping functionalities include *security* and *privacy* – if data associated with one functionality should not be available to another functionality; and *modifiability* – if a functionality will change, or will be modified by different people.

One technique that we use to systematically examine the properties that lead to coupling and cohesion is the *data coupling diagram*. Potential groupings are then evaluated and possibly refined using an *agent acquaintance diagram*. We explain each of these in detail in the following sections.

DATA COUPLING DIAGRAMS

One strong reason for grouping functionalities together is data coupling – they use the same data. In order to more easily visualize groupings suggested by data coupling, we can use the functionality descriptors to develop what we call *data coupling diagrams*.

A data coupling diagram consists of the functionalities and all identified data (not only persistent data but also data that the functionalities require to fulfil their job). Directed links are then inserted between functionalities and data, where an arrow pointing towards the data indicates the data is *produced* or *written* by that functionality, whereas an arrow pointing towards the functionality indicates the data is *used* by the functionality. A double-headed arrow indicates that the functionality both *uses* and *produces* the data.

The diagram must be checked to ensure that all data is produced somewhere, unless it has been determined that it is provided externally on system start-up and is static.

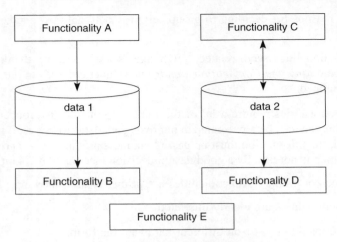

Figure 5.2   Simple data coupling diagram

Also, the existence of data that is produced but not used usually indicates an omission elsewhere in the design.

Once the diagram is complete, it can then be assessed visually for groupings that are linked by their data use. When assessing the diagram visually, we are looking for clusters of functionalities around data. Each such grouping must also be assessed for cohesion – do the functionalities fit naturally together into a single area of responsibility? It must also be assessed with respect to whether there is some reason to keep the functionalities separated.

> ☞ TIP: Checking that all data is produced/written by some functionality can often reveal missed 'data produced' fields in the functionality descriptors. This is where a functionality should produce certain data, but this fact has not been recorded in the design. This is a typical example of refinements and revisions that are commonly made to the system specification documents during the architectural design stage.

Consider the simple example data coupling diagram in Figure 5.2. We see that A and B exhibit data coupling as do C and D, while E is not data-coupled to anything. Let us assume that A, B, C and E are somewhat conceptually related, while C is also conceptually related to D. Let us also assume that C should not be in the same agent as A, as they need to be on different platforms. We tentatively put A and B into agent X, because of their data coupling, and C and D into agent Y. These decisions are consistent with cohesiveness, and also avoid the undesired grouping of C with A. Since E is conceptually related to A and B, it could be included into agent X without destroying cohesion. Alternatively, we could create a third agent, Z, containing functionality E. Consequently, we decide to consider two alternative groupings: {Agent X = A, B, E; Agent Y = C, D} or {Agent X = A, B; Agent Y = C, D; Agent Z = E}. Both these designs can then be considered further.

We note that the data coupling diagram does not capture all interaction between functionalities: some interaction may be based on message passing, possibly for transfer of control as well as of data. For instance, a sales transaction functionality may pass control to a delivery management functionality once it has completed its job. We have experimented using diagrams that show this kind of interaction, but have not found these as useful as the data coupling diagrams.

Electronic Bookstore: Case study

Taking the functionality descriptors for the electronic bookstore (see Appendix A) and drawing a data coupling diagram, we obtain the figure below:

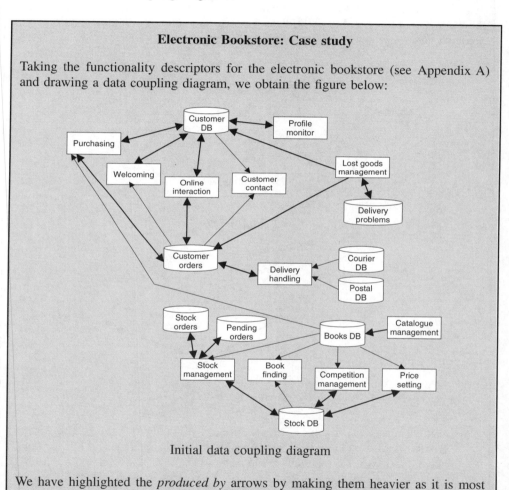

Initial data coupling diagram

We have highlighted the *produced by* arrows by making them heavier as it is most critical to keep writing of a data source within a single agent.

Sometimes, as in the bookstore example, there are initially no clear groupings that emerge from the data coupling diagram. In such a case, it is necessary to then make some design decisions that enable groupings to be made. These may include such things as deciding that a particular functionality will manage all interactions (or at least all writing) with a particular data store, or rethinking the details of scenarios and functionalities to eliminate some of the complexity of data interactions.

Electronic Bookstore: Case study

In the initial bookstore data coupling diagram (shown previously), we observe that there are many *produced by* arrows from different functionalities for the data sources *Customer DB* and *Customer Orders*. We make the design decision that all writes to these data sources will go via the functionalities *Profile Monitor* and *Delivery Handling*. We remove all *produced by* arrows to these data sources, other than from these functionalities. At this point, there emerge three clusters of functionalities (i.e. potential agents) as shown in the following figure.

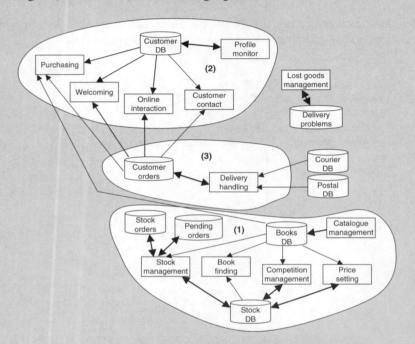

Natural functionality groupings after removing write links

- The first cluster (1) involves *Stock management* and *Price setting*. The Stock DB is written to by both *Price setting* and *Stock management* and so we want to keep the two functionalities in the same agent. The book finding functionality reads both the data stores here and so we (for now) put it in the same agent.

- The second cluster (2) revolves around the *Customer DB* and interactions with the customer. These functionalities read and write the *Customer DB* and some also read *Customer Orders*.

- The third cluster (3) is the functionality and data stores associated with delivering orders. Note that we have moved the data stores

> *Courier DB* and *Postal DB* outside the cluster as we have previously determined that these will be external data stores.
>
> - The functionality for *Lost goods management* stands alone at this stage and could be incorporated into one of the three clusters.

We have indicated previously a number of reasons for and against grouping particular functionalities together, and these must all be considered in relation to the initial groupings suggested by the data coupling diagram. Depending on the situation, these considerations may even result in design decisions being made prior to the initial data coupling diagram. For example, if it is known that certain functionalities are required to reside on a separate hardware platform, a decision may be made early on as to which data stores are required on that platform. Two data coupling diagrams may then be developed, one for each platform, to facilitate further decisions about groupings into agents.

Cardinality is one of the key issues to consider, and this is addressed for the bookstore example that follows.

Electronic Bookstore: Case study

In analysing the grouping so far with respect to cardinality issues, we note that the functionalities *Online interaction*, *Welcoming* and *Purchasing* all require one instance per customer, whereas *Profile monitor* and *Customer contact* require one instance for the whole system. Consequently, we split these into two separate groupings as follows. Note that when splitting the clusters, data is kept with the functionalities that produce it.

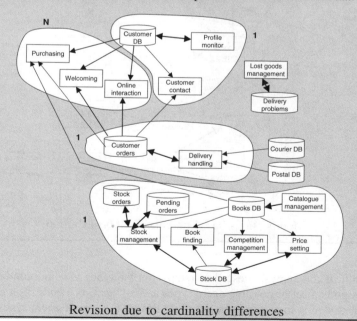

Revision due to cardinality differences

As has been indicated previously, *cohesion* is extremely important, and whatever clusterings are being considered, cohesion of the resulting groupings is a major concern. Adjustments to groupings may well need to be made or considered in order to improve this design aspect.

Electronic Bookstore: Case study

The first issue we look at in reviewing cohesion is whether the *Lost goods management* functionality can sensibly be placed with any of the existing clusters. We determine that it belongs well with *Delivery handling*, in a grouping that deals with deliveries of customer orders.

We also consider whether *Book finding* may fit better, in terms of cohesion, with the *Online interaction* grouping than with the *Stock management* one. The next figure shows both the above changes. We decide to explore two possible agent groupings, with the two alternative placements of *Book finding*.

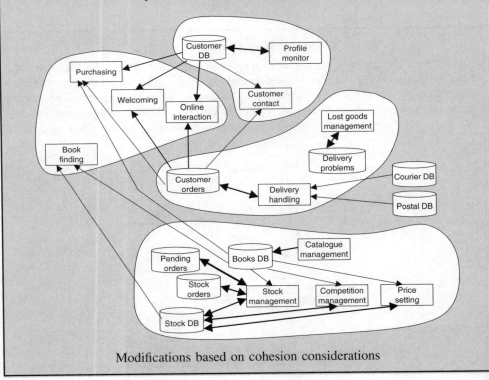

Modifications based on cohesion considerations

The result of the analysis described (step 1 on page 56) is some number of designs where a design is a set of possible agent types and an agent is a collection of functionalities. The next step is to assess each proposed design (i.e. set of agent types) and decide on the most appropriate one.

☞ TIP: A simple but well-used and effective heuristic for assessing cohesiveness is whether a suitable name for the agent can be found that adequately encompasses all of its functionalities. An agent that is cohesive should be able to be described by a single term without any conjunctions ('and'). For example, *Sales Assistant* agent is a simple descriptive name suitable for the combination of the functionalities of *Book Finding*, *Purchasing*, *Welcoming* and *Online interaction*. This suggests that the grouping of functionalities is cohesive. While only a 'rule of thumb', this is often an effective (and quick) way of assessing cohesion in proposed designs.

5.3 REVIEW AGENT COUPLING – ACQUAINTANCE DIAGRAMS

One of our design criteria is to aim for a system that is as loosely coupled as possible. We do not want all agents to have to know about all other agents.

In order to evaluate a potential grouping for coupling, we use an *agent acquaintance diagram*. This diagram represents each of the agent types. Information about agent interaction is extracted from the functionality descriptors and each agent is linked with the other agents it interacts with. Links can be decorated with the cardinality of the relationship if desired (e.g. one *Stock Manager* agent interacts with many *Sales Assistant* agents).

We then analyse the resulting diagram in two ways. One is simply an analysis of the density of the links within the diagram. It is a measure of the ratio of the actual coupling to the maximal possible coupling. If the system has four agents, then each agent could potentially be linked to a maximum of three other agents, giving a total number of $4*3/2$ possible links.[2] To get the link density, we simply count the links and divide by this number.

It is likely that all agents in the system are linked in some way, and if this is the case, the minimum number of links will be one less than the number of agents. An obvious design that has such a number of links is a star design, where each agent is coupled to one central agent. Although this has a low level of coupling, it is an undesirable design for two reasons: firstly, change to the central agent type is likely to affect many of the other agents; and secondly, it is a situation that can potentially lead to a bottleneck at run time when many agents are trying to communicate with a single agent.

Consequently, we also consider bottlenecks. If we ignore agent cardinality, the bottleneck factor can be assessed simply by examining the number of links that an agent has. The largest number of links from an agent type can be used as an indication of the worst bottleneck in the system for the purpose of comparing designs.

---

[2]Links are not directional, so a link from A to B is the same as a link from B to A, hence the division by two.

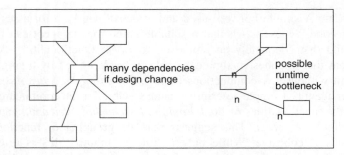

Figure 5.3   It is important to consider different kinds of bottlenecks in the design

However, it is also important to consider situations in which there are some larger numbers of agent instances of a given type as this can produce run-time bottlenecks (Figure 5.3). For example, if we have multiple *Sales Assistant* agent (one per customer) communicating with a single *Stock Manager* agent, this is a potential run-time bottleneck. To decide whether it is actually a problem requires consideration of the actual situation – how many customers do we expect to be dealing with concurrently, and how many interactions do we expect the *Sales Assistant* agent to have with the *Stock Manager* agent? It is not possible to provide a metric for assessing bottlenecks, partly as the number of instances is often not fixed, but rather is a function of system usage – for example, one *Sales Assistant* agent for each customer. Also, the numbers themselves do not necessarily provide useful information. Rather, it is important to consider the possible bottleneck issues and weigh these against the various other factors in deciding the preferred design.

A design with lower link density is less highly coupled and is therefore preferable, all else being equal. However, issues such as bottlenecks and cohesion, as well as possible additional factors such as agent size, must all be considered.

Electronic Bookstore: Case study

The agent acquaintance diagram is identical for each of the two groupings to be considered in our electronic bookstore example and is shown below:

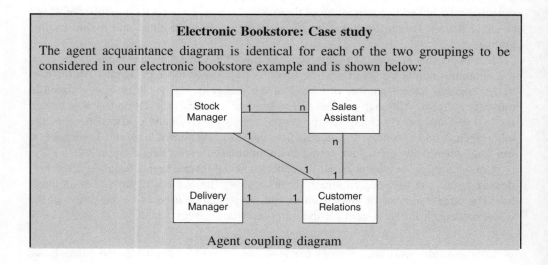

Agent coupling diagram

We note that this configuration is quite highly coupled, with a link density of $\frac{4}{6}$. We also observe that there are potential run-time bottlenecks with multiple *Sales Assistant* agents talking to one of the other agents. However, on consideration it is not felt that the number of *Sales Assistant* agents or the volume of communication will be a problem.

Possibly, further design work could lead to a more decoupled design that was also cohesive and met all constraints. However, we decide to continue with the design that incorporates *Book finding* into the *Sales Assistant* agent.

☞ TIP: The agent acquaintance diagram can also be used to review whether all the links are in fact necessary. Often, in reviewing an agent acquaintance diagram some links seem natural – you would expect those agents to interact – whereas others appear less intuitive. It is often possible to eliminate some of the less intuitive interactions by revisiting the scenarios and revising ideas as to what should happen when. This can often lead to a cleaner, more decoupled design than was initially possible.

5.4 DEVELOP AGENT DESCRIPTORS

The decision as to which design is to be pursued is made after reviewing the agent acquaintance diagrams and considering issues such as agent size and cohesion.

During and after making the decision as to which agent types we will have within the system, there are a number of questions that need to be resolved:

- How many agents of each type will there be (singleton, a set number, or multiple instances based on dynamics of the system, for example one *Sales Assistant* agent per customer that arrives at a website)?

- What is the lifetime of each agent? If agents are created or destroyed during normal system operation (i.e. other than at start-up and shut-down of the whole system), what triggers this? For example, a new agent might be created when a user accesses the system for the first time.

- Agent initialization – what needs to be done?

- Agent demise – what clean up needs to be done?

Each agent type should have an agent descriptor containing the just-mentioned information plus the *name* of the agent, a natural language *description* of what this agent does within the system and a list of the *functionalities* from the previous phase that are incorporated within this agent.

In addition, we extract from the functionality descriptors the following information that becomes a part of the agent descriptor:

- What are the goals of this agent?

- What percepts will this agent react to?

- What actions (if any) will it take?

- What data does this agent use or produce?

We also require a brief description of the *interaction protocols* that it uses.

Additional details are also added during *Detailed Design*. The complete descriptor can be seen in Appendix A.

Electronic Bookstore: Case study

For example, consider the following agent descriptor from our electronic bookstore example:

Name: Sales Assistant agent.
Description: Greets customer, follows through site, arranges purchases.
Cardinality: One/customer.
Lifetime: Instantiated on customer arrival at site. Demise when customer logs out or after inactivity period.
Initialization: Obtains cookie. Reads Customer DB, Stock DB.
Demise: Closes open DB connections.
Functionalities included: Online interaction, Purchasing, Welcomer, Book finding.
Uses data: Customer DB, Customer Orders, Books DB, Stock DB.
Produces data: Customer preferences, orders, queries
Goals: Welcome customer; Update customer details; Respond to queries; Facilitate purchases;
Percepts responded to: *Arrival at WWW site, Bank transaction response.*
Actions: *WWW page display, Bank transaction*
Protocols and interactions: *Book finding* with *Stock Manager*, *Book ordering* and *Order status querying* with *Delivery Manager*.

6

Architectural Design: Specifying the Interactions

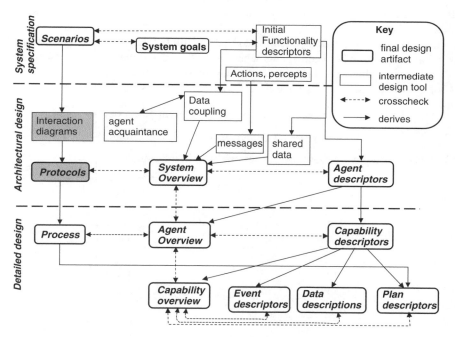

Phases, artifacts and relationships in the design process

Once the agent types are decided, the next aspect of the architectural design is to specify the interaction between agents, capturing the *dynamic* aspects of the system. This builds on both the agent types and also on the scenario descriptors from the System Specification phase.

Developing Intelligent Agent Systems   L. Padgham & M. Winikoff
© 2004 John Wiley & Sons, Ltd   ISBN: 0-470-86120-7 (HB)

The development of the specification of the interaction between agents (step 2 in Figure 5.1) has the following stages:

1. Develop interaction diagrams from use case scenarios.

2. Generalize interaction diagrams to interaction protocols.

3. Develop protocol and message descriptors.

Like use case scenarios, interaction diagrams describe only a single example of the system's behaviour. The fully specified *interaction protocols* are then developed from these and are the final design artifact.

6.1 INTERACTION DIAGRAMS FROM SCENARIOS

Interaction diagrams are borrowed directly from object-oriented design, but show interaction between agents rather than objects. The process for developing interaction diagrams is to take the scenarios developed in the specification phase and to build corresponding interaction diagrams.

This involves (a) replacing each functionality with the agent that includes it; (b) inserting a communication between agents where it is needed and then (c) expressing the result as an interaction diagram. This process is summarized in Figure 6.1.

For example, suppose we have a scenario that includes the following steps:

1. GOAL: Update user record (functionality Student Records)

2. GOAL: Check money owing (functionality Student Finances)

3. GOAL: Request graduation (functionality Graduation)

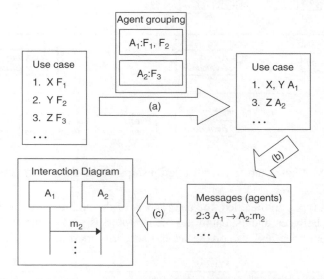

Figure 6.1   From use case scenarios to interaction diagrams

Suppose that the first two functionalities (Student Records and Student Finances) are both in an agent 'Student Manager' and that Graduation is in an agent 'Awards'. Then the first two steps are done by Student Manager and the third step by Awards. The Awards agent needs to be told (by Student Manager) when to request a graduation, thus there is a message from Student Manager to Awards after step 2 (and before step 3).

Of the three steps depicted in Figure 6.1, the first and third are quite simple. Step (a) involves replacing functionalities with agents on the basis of the grouping of functionalities into agents. If a functionality F has been grouped into agent type A, then steps in a use case scenario that are done by the functionality F are considered to be done by the agent A.

The result of step (b) is a list of messages. Step (c) involves a simple change of notation from this list of messages (with sender, recipient and message details) to the interaction diagram notation.

☞ DEFINITION: **Interaction Diagram Notation:** The notation used for interaction diagrams is standard. Time increases as one moves down the page. Each agent has a *lifeline*, depicted as a vertical line with the name of the agent in a box at the top of the line. Messages are depicted as horizontal arrows between lifelines with a brief description of the message above the arrow. The simple example below shows two agents named *User* and *System* where the User sends a Query message to the System. This is followed by the System sending a Response message to the User.

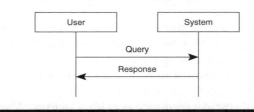

☞ TIP: **Depicting Actions and Percepts:** Sometimes, showing actions and percepts on interaction diagrams can be important. For example, often an interaction is triggered by a percept. There are a number of ways that actions and percepts can be depicted in interaction diagrams. They can be shown as messages from an invisible lifeline (left in the example below), they can be shown as text on the lifeline of the relevant agent (centre in the example below), or they can be shown as messages to an explicit environment lifeline (right-hand side, below).

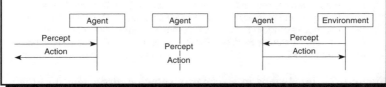

In the remainder of this section, we focus on the interesting step: determining where communication between agents is required. Wherever there is a step in the scenario that involves a functionality from a new agent, there must be some interaction from a previously involved agent to the newly participating agent. While it is not possible to automatically derive the interaction diagrams from the use case scenarios, substantial consistency checking is possible (see Section 4.5).

More generally, determining when communication between agents is needed (step (b) above) involves identifying the sequential dependencies between steps in the scenario. Whenever a step n must occur after step m (and the two steps are performed by functionalities that are not in the same agent), then there needs to be a message.

Analysing the sequential dependencies will identify where messages are required, and we then need to name the message type, and eventually develop a message descriptor for the type. Initially, it is usually useful to assume that each message in an interaction diagram (or protocol) is a new type (i.e. has a unique name). It may later be possible to coalesce some number of messages into a single implementation type, but for clarity of design information, they should usually be kept separate at this stage.

In many use case scenarios, the steps are sequential: step 2 must occur after step 1, step 3 must occur after step 2, and so on. However, this is not always the case. For example, looking at Figure 6.2 showing the end of the scenario where a user buys a book (Order Book scenario), the last three steps involve the Stock Manager logging the outgoing delivery (step 12 below), the customer relations agent registering the purchase in the user's profile (step 13) and sending the customer an e-mail (step 14). Although the use case scenario presents these steps in a particular order, in fact they are not required to occur in that order.

Looking at Figure 6.2, we see that steps 10 and 11 are in the same agent (actually in the same functionality), so no message is required. We now consider a number of possible dependencies in the remaining steps.

Option 1: The steps are required to be in the given sequence, that is, step 12 must occur after step 11, step 13 after 12 and step 14 after step 13.

In this case, we would clearly need messages from Delivery Manager to Stock Manager, and Stock Manager to Customer Relations. We might consider that as steps 13 and 14 are within the same agent, no message is required here. This would require the Profile Monitor functionality to tell the Customer Contact functionality to e-mail the customer when it has finished registering the purchase. However, this design decision would not be a good one: this scenario is dealing with ordering books, and the Profile Monitor should not be responsible for telling the Customer Contact functionality that it

	Step	Functionality	Agent
⋮			
10.	Place delivery request	Delivery Handling	Delivery Manager
11.	Log outgoing delivery	Delivery Handling	Delivery Manager
12.	Log books outgoing	Stock management	Stock Manager
13.	Update Customer Profile	Profile Monitor	Customer Relations
14.	Send email	Customer Contact	Customer Relations

Figure 6.2 Final steps in a scenario, showing also agent information

should let the user know that their order has been sent. Consequently, we require the Delivery Manager to notify the Customer Relations agent when it is time to send e-mail to the customer. This requires the addition of a reply message to the Delivery Manager, indicating when the profile updating is completed, as is shown in Figure 6.3.

Option 2: Steps 12, 13 and 14 must all occur after step 11, but can occur in any order. In this case, after step 11, messages are sent from the Delivery Manager to the other two agents, as shown in Figure 6.4 Because the interaction diagram captures a particular ordering, showing of this potential parallelism is left for the interaction protocol. The interaction diagram simply captures the same (arbitrary) ordering as that shown in the use case scenario. However, establishing whether there is a need for strict sequencing can effect what messages are necessary, as we have seen.

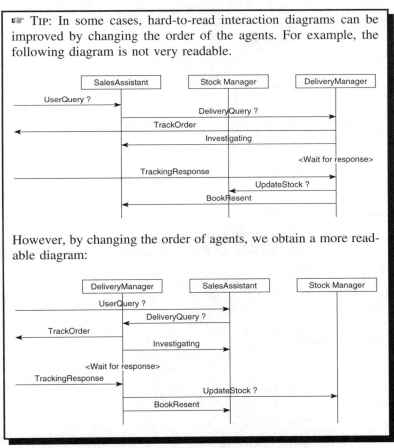

☞ Tɪᴘ: In some cases, hard-to-read interaction diagrams can be improved by changing the order of the agents. For example, the following diagram is not very readable.

However, by changing the order of agents, we obtain a more readable diagram:

☞ Tɪᴘ: **Including the trigger:** The table of steps in the use case scenario may not include the trigger. When developing the interaction diagrams, consider whether it is useful to include the triggering message or percept within the interaction diagram.

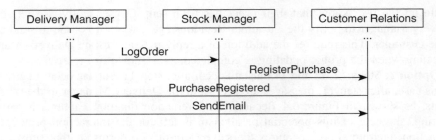

Figure 6.3   Option 1, interaction diagram for sequential steps in scenario

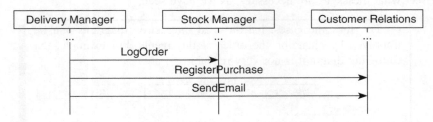

Figure 6.4   Option 2, interaction diagram where strict sequencing from scenario was not necessary

Following is a larger worked example from the Electronic Bookstore.

Electronic Bookstore: Case study

Here is the outline of the **Query Late Books Scenario** developed in the first round of scenario descriptions:

Query Late Books Scenario
Trigger: User query (via Online Interaction)

Key for functionalities:

DH   Delivery Handling
LGM Lost Goods Management
CC   Customer Contact
SM   Stock Management

	Step type	Step	Functionality
1	GOAL:	*Determine delivery status*	DH
2	GOAL:	*Log delivery problem*	LGM
3	ACTION:	*Request delivery tracking*	LGM
4	GOAL:	*Inform customer*	CC

5	OTHER:	*Wait for response*	DH
6	PERCEPT:	*Tracking info received*	DH
7	GOAL:	*Arrange delivery*	DH
8	GOAL:	*Log books outgoing*	SM
9	GOAL:	*Inform customer*	CC
10	GOAL:	*Update delivery problem*	LGM

We begin by replacing functionalities with agents. We use numbers in brackets to indicate steps that occur in the same agent as the preceding step.

Key for agents:

DM: Delivery Manager
CR: Customer Relations
SM: Stock Manager

	Step type	Step	Agent
1	GOAL:	*Determine delivery status*	DM
(2)		*Log delivery problem*	DM
(3)		*Request delivery tracking*	DM
4	GOAL:	*Inform customer*	CR
5	delay ...		
6	PERCEPT:	*Tracking info received*	DM
(7)		*Arrange delivery*	DM
8	GOAL:	*Log books outgoing*	WM
9	GOAL:	*Inform customer*	CR
10	GOAL:	*Update delivery problem*	DM

We now consider the dependencies between these steps. Steps 1 to 3 are within an agent, and so we do not capture the dependencies between them at this level.

Step 4, informing the user that tracking has been requested, can be done in parallel with steps 1 to 3, or could be required to be done only after the delivery tracking has actually been requested. In this case, we assume the latter: the system will wait until tracking has been requested before telling the user that it is investigating. By showing step 3 on the interaction diagram, we can show that step 4 must occur after the tracking is initiated.

Steps 5 (delay) and 6 (percept being received) must occur in sequence after step 4, and are thus shown in that order on the interaction diagram.

Step 7 (arrange delivery) is required to occur after step 6, but since it is internal to the Delivery Manager agent (which has just received the percept in step 6), no inter-agent communication is required.

Step 8 (log books outgoing) is required to be after step 7 (arrange delivery), because we do not want to log books as being sent until we are certain that the delivery has been arranged and they will be sent. So, once delivery has been arranged (which is not shown in the interaction diagram), the Delivery Manager sends a message (*Update Stock*) to the Stock Manager, at which point the Stock Manager can log the outgoing books.

Step 9 (inform customer) is required to be after step 7 (but not necessarily after step 8): we do not want to tell the customer that their books have been resent until we know that they actually will be sent (i.e. after step 7), but we do not need to wait for the stock to be updated in order to tell the customer, so step 7 and step 8 can occur in parallel. Thus, after step 7 there the Delivery Manager sends a message (*Book Resent*) to the Customer Relations.

The two messages (triggering steps 8 and 9) can be sent simultaneously.

Finally, step 10 is required to be after step 7: we do not need to wait for the customer to be informed or for the stock records to be updated in order to update the delivery problem record, but we do want to wait for the delivery to be arranged. So, step 10 must come after step 7; since both steps are performed by the Delivery Manager, the dependency is internal and no message is required between agents.

This yields the following interaction diagram. Note that the trigger of the scenario is a *Delivery Query* message (possibly from Customer Relations or Sales Assistant) received by the Delivery Manager.

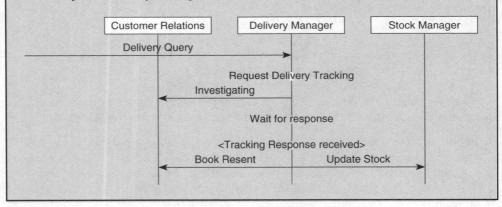

6.2 INTERACTION PROTOCOLS FROM INTERACTION DIAGRAMS

As with scenarios, we would expect to have only a representative set of interaction diagrams, not a complete set. In order to have complete and precisely defined interactions,

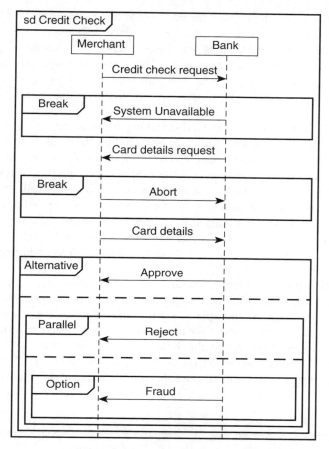

Figure 6.5   Protocol diagram for credit checking

we progress from interaction diagrams to protocols that define exactly which interaction sequences are valid within the system. Because protocols must show all variations, they are often larger than the corresponding interaction diagram and may need to be split into smaller chunks.

In order to give an example of developing a complete interaction protocol, we need a notation for describing interaction protocols. Since interaction protocols include sequencing, choices, iteration and other control structures, the notation needs to be quite expressive. A range of notations exist for describing protocols including UML activity diagrams, AUML (Agent UML) (Odell et al. 2000), and Petri nets (e.g. (Cost et al. 1999; Nowostawski et al. 2001; Poutakidis et al. 2002)). We will be using the revised version of Agent UML, (Huget et al. 2003) currently under development.

Developing protocols is done by considering alternatives. For each message (or percept) that an agent receives, we ask 'what are the possible messages that the agent could send as a response?' We then repeat the process for these messages. More generally, we ask 'what are the possible continuations (i.e. *sequences* of messages)?'

In some cases, the alternatives are best described in terms of parallelism or looping. For example, if an interaction diagram shows message m_1 followed by m_2 and an alternative is that m_2 is sent first followed by m_1, then we could describe this as an alternative, but it is simpler to say that the two messages are sent in parallel.

For example, consider a credit card–checking protocol involving a merchant and a bank. The protocol starts with the merchant sending the bank a credit card request message. The bank will respond to this with a card details request; there are no alternative responses. The merchant, upon receiving the card details request, will respond with the card details; again there are no alternative responses. When the bank receives the card details, it can respond with a number of possible messages. One possibility is to respond with an *Approve*. Another possibility is to respond with a *Reject*. Finally, in some cases, the bank will respond with both a reject message (to the merchant) and also a *Fraud* message. Figure 6.5 shows this protocol. Note that sending *Reject* and *Fraud* can happen in either order or simultaneously.

In considering possible alternatives, one source of possibilities is the variation field of the use case scenarios. Another source of possibilities is different use case scenarios. Often, there will be multiple scenarios that relate to the same concepts. For example, one scenario might show what happens when a customer orders a book that is in stock and another scenario might show what happens when a customer orders an out-of-stock book. Since interaction protocols are meant to cover all possibilities, we would consider both of these use case scenarios to be alternatives in a single interaction protocol.

Electronic Bookstore: Case study

The diagram below shows the protocol for the query late books use case scenario.

In developing it from the interaction diagram given earlier, we considered at each step what were the possible alternatives. In this case, we focus on the alternatives that occur in response to the tracking request. The use case scenario given in Chapter 4 outlines three variations to the case where the book is determined to be lost and is resent:

1. The book is within (or close to) the time window of expected delivery. In this case, no request is issued to track the order and the user is informed that the order is expected soon.

2. The tracking information received indicates that the book is on its way, thus there is no need for a new delivery.

3. The tracking request does not return any information (*NoTrackingResponse*). In this case, the book is resent.

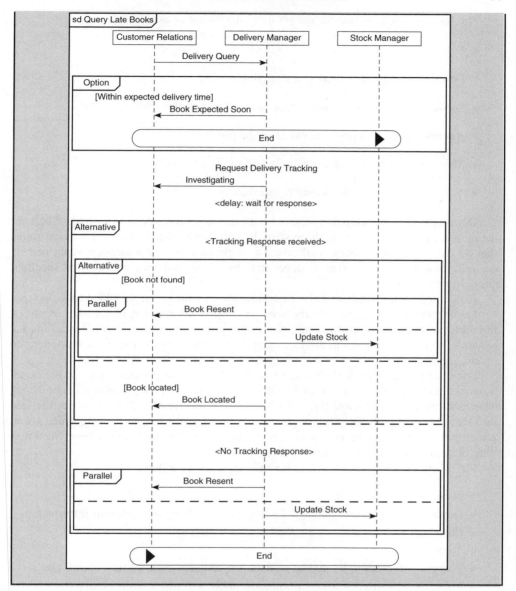

6.3 DEVELOP PROTOCOL AND MESSAGE DESCRIPTORS

As with each of the final design entities in the system, we have descriptors for protocols and for messages. Descriptors allow us both to collect together information existing in other places and also to specify additional details about the particular entity.

The template that we use for the protocol descriptor is as follows:

- Name

- Description – brief natural language description

- Scenarios – lists the scenarios that are included in this protocol

- Agents – lists the agents involved in the protocol

- Messages – lists the messages involved in the protocol

- Notes – and other miscellaneous information.

During protocol development, a number of messages have been identified, which are the messages between agents in the system. In some cases, there are inter-agent messages that do not require a protocol, as they are a single message, or a message reply pair. If desired, these can be specified as degenerate protocols, but they can also be specified simply as messages.

In detailed design, there will also be additional *internal messages* added to the design.

Descriptors for messages clearly need to contain information about the content of the message. It is also useful to indicate the purpose of the message, which may be, for example, to transfer control to another agent or functionality, to request a service, or to update information.

For example, a Sales Assistant agent in the Electronic Bookstore may tell the Stock Manager that a book has been sold, with the primary purpose of updating the Stock Manager's records. The same Sales Assistant agent may send a message about the sale to a cashier agent with the primary purpose of transferring control to the cashier agent, which will then obtain necessary details for doing the financial transaction. Messages may of course perform multiple functions.

Fields in our template descriptor for messages are as follows:

- Name – brief identifier
- Description – natural language description containing any relevant information
- From agent
- To agent
- Purpose – for example, update of data, request service, transfer control, and so on
- Information carried – what information fields will this message carry.

Some additional fields will be described and added during detailed design.

Electronic Bookstore: Case study

An example message from the Electronic Bookstore is *Order Sent*, which is sent by the Delivery Manager agent to the Customer Relations agent when a pending order has been sent out.

- **Name** – Order Sent.

- **Description** – contains information about the contents of a pending order that has now been sent. It allows the sending of notification to the customer and updating of the customer record.

- **From agent** – Delivery Manager.

- **To agent** – Customer Relations.

- **Purpose** – to update the customer record and also to trigger sending an e-mail to the customer, if appropriate.

- **Information carried** – order reference, list of books sent, customer ID, date, carrier.

7

Finalizing the Architectural Design

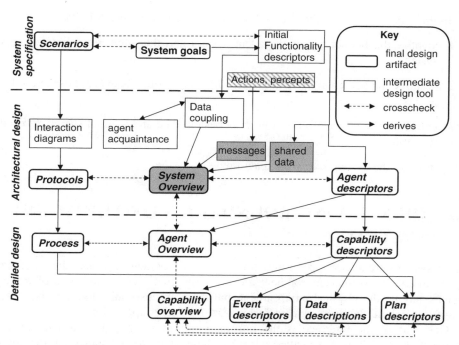

Phases, artifacts and relationships in the design process

Having specified the agents within the system and the communications between them, we now bring this information together in a *System Overview Diagram*, which captures diagrammatically the overall architecture of the agent system. In order to do this, we also need to develop a fuller understanding of the data needed as well as of the environmental

Developing Intelligent Agent Systems L. Padgham & M. Winikoff
© 2004 John Wiley & Sons, Ltd ISBN: 0-470-86120-7 (HB)

information that will be responded to and the actions that the agent may take within the environment.

7.1 OVERALL SYSTEM STRUCTURE

In building a multi-agent system, it is often the case that some portions of the overall system are more appropriately developed outside of an agent paradigm. Such aspects typically fall into one of four categories:

- Data representation and management

- Percept processing

- Action management

- Pre-existing code.

In this final step of the System architecture phase (see Figure 5.1, Page 54), we

1. identify the boundaries of the agent system and the interactions with other sub-systems;

2. describe the percepts and actions, and the relationships between these and relevant agents;

3. define all shared data, both external persistent data and internal shared data within the system;

4. develop the system overview diagram.

7.2 IDENTIFYING BOUNDARIES OF THE AGENT SYSTEM

Agent systems are always operating within some environment, and thus the specification of the interfaces to that environment are very important. The boundaries of the agent system may be the physical environment, but there are also quite likely to be software boundaries where the agent system forms part of a larger application.

One part of the interface is the information coming from the (usually dynamic) environment, which we have called *percepts*. Depending on the application, the raw data from the environment may require significant processing in order to be of any use to the agents in the system. It may well be more appropriate for this processing to take place in a non-agent sub-system, than to incorporate it within an agent. Reasons for this may include existence of more suitable languages than that being used for agent development, lack of benefit of the agent paradigm or pre-existing libraries that can be used.

For example, in an agent system controlling a robot, the perceptual input may be vision data from a camera, in the form of a series of bitmaps. However, this requires significant pre-processing before it is of interest to the agent. Firstly, there must be some processing to identify relevant objects within the bitmap. However, even this is not really sufficient. In order to act within the environment, the agent wants to notice when interesting or significant things have happened. It does not want to reason about every frame that is essentially the same as the previous one. Typically, it will want to notice differences between frames, particularly unexpected differences. It may also notice differences between frames and its expectations – for example, if an agent believes, on the basis of previous information, that a ball is just to its left, but when it turns and looks at that spot it does not see a ball, then this is significant information.

We use the term percept to refer to the environmental information that the agent receives, although we acknowledge that this may well have received significant processing from the raw data obtained.

Similarly, there may be some aspects of managing the agent's actions in the environment, which are best performed outside of an agent paradigm. Using our robot example again, a typical action may well be to move to a specified point. However, there is a significant amount of motor control programming required in order to effectively implement this. This is most appropriately seen as a behavioural sub-system for each agent, which is designed and implemented outside of the agent paradigm being described here.

It may also be the case that pre-existing code performing some relevant functionality is to be incorporated into the agent system. The interface between such sub-systems and the agent system must also be identified. This can be represented in terms of messages, which may be requests and responses, but may also include incoming messages generated by the external sub-system if it runs asynchronously and can generate data other than as a synchronized response to a request. Alternatively, the external sub-system can be seen as part of the environment, and interaction can be represented in terms of percepts and actions.

The implementation of the interface to an external sub-system may involve wrapping pre-existing code as agents and using agent messaging, or may simply be the use of a defined API. This decision is not important at design time and is left as an implementation decision. However, the information needed and provided by the pre-existing sub-system is important and should be documented as part of either messages or percepts and actions.

Issues to Consider

Some issues to consider when making the decision regarding whether to include something within the agent system or to make it a separate sub-system are as follows:

1. Is there any benefit (currently or in future extensions of the system) of the agent paradigm with aspects such as autonomy, flexibility and goal orientation?

2. Does keeping it separate make for less coupling, allowing it to be developed separately?

3. What is the simplest approach to integrating, or interfacing to, existing code?

4. Does the processing in question need to be on a different time cycle than the agent processing? For example, processing frames from a video camera may need to be done 50 times a second, whereas higher-level decision-making may take place at a much lower rate (such as five times a second).

7.3 DESCRIBING PERCEPTS AND ACTIONS

In this step, we develop descriptors for the input and output interface entities, percepts and actions, and link these to relevant agents.

Percepts are the information (possibly processed) that the agent receives from the environment, while actions represent the effects the agent can have on the external environment. We discuss these, identifying issues to consider and providing a template for recording of important information for understanding the design.

> ☞ TIP: Although the exact details of the interface to the external environment may be considered as implementation, or perhaps detailed design, it is critical to establish some aspects of this at an early stage. If it is impossible, or very difficult, to obtain particular data, then this will affect the design of the system at a high level.

PERCEPTS

It is important at this stage to clarify exactly what information is provided as dynamic input to the agent system. Most percepts will be a result of ongoing events happening in the external environment, but some may also have to do with initialization of the agent system. For example, a percept 'TCP/IP connection established' may be something that will be noticed by the agent at system start-up and which should cause some agents to read in data and establish internal data structures.

Identification of percepts to be obtained is driven by the information the agent system requires in order to achieve the goals identified in the initial phase. However, this is also constrained by what is feasible given the sensors or other mechanisms for obtaining data and the processing techniques available. Where substantial processing of raw data is required, great care should be taken to ensure that extraction of relevant information is feasible, prior to continuing the design of the agent system.

It is important to identify the format of the raw data, including whether there are any important intermediate formats of relevance for the pre-processing and extraction of agent percepts. For example, the vision frame from the robot camera has an initial format as a bitmap of a certain size. An intermediate format is a data structure containing certain objects. If existing software is being used, this data structure may have pre-defined contents or limitations, which are important to understand. It is then important to define the information contained in the final processed percept, as well as any additional information that may be required to obtain the desired percept.

For example, suppose we want to extract a percept 'ball at position X, Y' from vision data, with an intermediate representation that provides objects and their global location information. We want this percept generated only when it is new information, not at every frame. In order for this to happen, we must at least maintain information regarding (one or more) previous frames. In addition, we may need access to information regarding what the agent currently believes – if it has momentarily turned so that the ball is out of sight, then turns back, it may be inappropriate to generate a percept when the ball is perceived in its expected location, despite the fact that it was not seen in the previous frame. On the other hand, failure to see the ball as expected may require percept generation.

It is also important to consider the likely frequency of such percepts and whether this is manageable by the agent system. If, for example, it is possible that a certain type of percept may on occasion happen with too high a frequency to be managed, then this may be a reason for some pre-processing. The percept could in normal cases be passed on directly, but in cases of high frequency we might need to either pass on only some percentage of the percepts, or aggregate percepts before passing them on.

Many percepts will result in an update of the agent's knowledge as well as potential action on the part of the agent. Knowledge updates resulting from the percept should be explicitly identified.

Our template descriptor for percepts is as follows:

- Name – Short name suitable for use in diagrams.

- Description – Brief description indicating the situation in which this percept is received.

- Information carried – Indicates what information is available regarding this percept. This often requires careful checking regarding the actual interface to the external environment.

- Knowledge updated – This will be either directly extracted from the information carried, or will be a result of some reasoning about the information carried.

- Source – Indicates how the information is obtained from the environment. In physical systems, this is likely to be a sensor.

- Processing – Indicates whether the percept is received directly from the agent sensor, or whether it is first processed in software, and if so how.

- Agents responding – Indicates which agents react to this percept in some way.

- Expected frequency – Describes how often one would expect to receive percepts of this type, with additional information regarding handling of extreme cases if appropriate.

Electronic Bookstore: Case study

An example of a percept from the electronic bookstore example is *Failed Stock Arrival*.

- **Name** – Failed Stock Arrival.

- **Description** – Occurs when stock is expected to arrive at a certain time, but the time passes (with some margin) and it does not arrive.

- **Information carried** – The reference to the stock order.

- **Knowledge updated** – Reliability of supplier, expected arrival date for waiting orders, expected arrival date for stock.

- **Source** – An indication from a previously set system alarm clock. (The stock orders DB is also accessed for information regarding the particular order.)

- **Processing** – When the 'alarm' goes off, records may need to be checked to ensure that the order has not in fact arrived, or been cancelled, or been previously notified.

- **Agents responding** – *Stock Manager* reacts to percept and notifies *Delivery Manager* if necessary, which may in turn notify *Customer Relations*.

- **Expected frequency** – Occasional.

> ☞ TIP: Percept processing is often an area where some prototyping is necessary at an early stage to avoid development of a design that is infeasible because of the difficulty of reliably extracting the required information from the raw data.

ACTIONS

Actions, like percepts, may also be complex, requiring significant design and development outside the realm of the reasoning system. This is especially true when manipulation of physical effectors is involved.

Actions can be classified depending on a number of properties:

- Is the action effectively instantaneous, or does it take a certain amount of time to perform (durational)?
  For example, sending an e-mail is an effectively instantaneous action. Following a ball is a durational action. Another durational action might be downloading a large file.

- If the action is durational, does the action have parameters that can be changed while it is executing?
  Downloading a file usually does not allow any parameters to be set – one simply waits for the download to complete. On the other hand, following a ball has a number of parameters including speed, the amount of acceleration that should be applied (for example, it might be desirable to conserve energy by using lower

acceleration levels), and so on. These factors apply to the design of the actions (outside of the agent part of the system) but also affect the design of agents, functionalities and plans that use these actions.

- Can the action fail? Most actions involving the real world can fail. Even in a software-only system, failures can occur: an e-mail address may be wrong, a web-site may be down, a network connection may be temporarily out of action.

- If the action can fail, does it always report failure when it occurs?

- Can the action have partial effects? For example, an e-mail is either delivered or not delivered. An attempt to move to a location may fail if the robot gets stuck halfway, but will still have changed the state of the world.

If actions are durational, that is, they take time, it is important to consider whether it is appropriate for the agent to do other tasks while this action is executing. In robot soccer, moving to a location is durational. While it is not appropriate for the agent to do other actions whilst doing this as it uses all the effectors, and any other action would interfere, it is important that the agent is free to do reasoning tasks on the basis of new perceptions.

In our electronic bookstore example, most of the actions are effectively instantaneous, such as sending an e-mail. Some, such as querying the status of a delivery, can be viewed either as a durational action that includes waiting for a reply, or as an instantaneous action of sending an e-mail (where receipt of the reply lies outside the action).

Our action descriptor template is as follows:

- Name

- Description – Indicates what the action is and its intended result.

- Parameters – Identifies parameters that can affect how the action is performed. For example, which printer to use may be a parameter to an action to print a file. Which speed to use may be a parameter for a robot move-to action.

- Temporality – Durational/instantaneous. If durational, indicate approximate length of time.

- Failure detection – Is there a notification if failure occurs (yes/no/maybe)? If so, what form does that notification take? If there is no notification, is it possible to check for success/failure in some way?

- Partial change – If the action does not succeed, what, if any, is the change that results from doing (or attempting) the action?

- Side effects – Indicates the side effects of an action, for example, the side effect of a move-to action may be a reduction in battery power.

Electronic Bookstore: Case study

An example action from the electronic bookstore is to order new stock:

- **Name** – Order stock
- **Description** – This action orders a list of books, in varying quantities, from a single supplier. The result of the action is that the supplier receives the order. An expected consequence is that the supplier sends a message confirming the order, and that eventually the books arrive. The latter are indirect results of the action rather than direct results.
- **Parameters** – Supplier e-mail address(es), order list, priority.
- **Temporality** – Instantaneous. (Indirect effects may take some days.)
- **Failure detection** – E-mail bounce may be detected. Only way to check success is to send e-mail requesting confirmation. Failure may go undetected.
- **Partial change** – Mail message file may exist.
- **Side effects** – None.

The percepts and actions provide the necessary interface between the system and its larger environment. To complete our understanding of the system at this level, we also need to include data sources.

7.4　DEFINING SHARED DATA OBJECTS

During the system specification stage, the persistent data has been identified. Now decisions must be made about what kind of persistent data stores will be used – a database or a file, and earlier ideas about which data items will be grouped together in a single data store must be confirmed and developed.

It must also be determined whether there will be any shared data within the system, accessible to more than one agent. A good design will minimize this, but there may be situations where it is reasonable to have shared data objects.

If multiple agents will be writing to shared data objects, this will require significant additional care for synchronization (as agents operate concurrently with each other). Consideration must be given as to whether this can potentially result in inconsistent data states containing parts of two different updates. This will depend partly on details of the implementation environment. Simple forms of locking can be used. We do not discuss locking further here since it is a standard topic in concurrent programming. However, it is important to ensure that locks are released both on success and failure.

☞ TIP: A common problem in multi-agent systems, which can be especially hard to debug, is *race conditions*. Data stores that are used by more than one agent can easily lead to race conditions because of synchronization issues. This can result in nondeterministic execution where the results of running the system can differ, even if the inputs are identical. This is one of many good reasons to avoid shared data stores.

A race condition is where the behaviour of a program depends on timing. Usually, such behaviour is unexpected. For example, consider two people, Alice and Bill, trying to update a bank account by reading the amount of money, adding 100 to it, and saving the result back to the account. The desired (and expected) result is that the account balance will be increased by 200. However, if the following sequence takes place, then the account will be incremented by only 100:

1. Alice reads the account balance (say, 150)

2. Bill reads the account balance (150)

3. Bill computes balance + 100 (i.e. $150 + 100 = 250$)

4. Bill saves balance + 100 (i.e. 250) to the account

5. Alice computes balance + 100, that is, $150 + 100$ (note that the original balance is being used, since that is what Alice read in step 1)

6. Alice saves balance + 100 (i.e. 250) to the account, overwriting Bill's change.

Database systems use transactions to avoid these problems, but such mechanisms are not always available or appropriate in agent systems.

Most applications have significant amounts of data with complex structure. If this data is to be stored within an external database, then it is appropriate to use database design techniques for specifying the design. If the data represents instances of interrelated classes that will be kept in main memory and possibly stored externally in some file format, then it is appropriate to use object-oriented design. The design should be developed and documented during this and subsequent phases, but we do not provide details here as there are already plenty of available materials providing methodology. Data may also have different internal and external representations.

Some important questions to ask at this stage about persistent data objects are as follows:

- Is there an in-memory version of the data?

- When is the external data updated – whenever an in-memory version changes, or not?

- Are any consistency checks needed on this data – if so what?

- How do agents access this data? (via requests to another agent, via DB access, reading a file, direct access to attributes of data object, via methods on a data object, and so on.)

It is important to note that when we talk about data, we need to differentiate data objects from data types. For example a hospital system may have a patient database which contains a set of patient records. There may also be a tuberculosis register which all hospitals in the state are required to use, also containing a set of patient records. It is not sufficient in the design to show these both as being of the same type, *patient records database*. To do so would have the effect of showing only one copy in design diagrams, which is very confusing, as they serve very different purposes. Rather, the two databases should be separated at design time, but would later be implemented as instances of the same class. Saying that we have a *tuberculosis database* and a *hospital patients database* provides useful information. Thus data stores that are collections of records are usually captured at the instance level on design diagrams, rather than at the type level as for other entities.

For shared data that is not persistent – that is, it exists only at run time, the main issues (other than the structure of the data), have to do with potential inconsistencies. If multiple agents can write data, what mechanisms prevent potential destruction of data? Are there any intermediate points when data is being modified, where it should not be accessed, due to incompleteness or inconsistency?

It is important to record with the documentation of the data, relevant information that is also available elsewhere in the design. This includes which agents read and write the data.

Our template for data descriptors is as follows:

- Name

- Description – What this data object represents.

- Data type – This should refer to part of a class diagram, or a database view.

- Included fields/aspects – What data, (perhaps named as necessary for certain functionalities) is considered part of this data object (or corresponding class).

- Produced by – Agent(s), capabilities, plans.

- Used by – Agent(s), capabilities, plans.

- Persistent – Indicates whether it is stored on an external device, file, database, etc.

- Initialization – What is initialization process or values.

- Used when – Indicates what the data is used for and when it is updated.

☞ TIP: Often, what at first appears to be a shared data object can be re-conceptualized as a data source managed by a single agent, with information provided to other agents as they need it. Alternatively, each agent may have its own version of the information, without there being any need for a single centralized data object.

7.5 SYSTEM OVERVIEW DIAGRAM

The above steps have identified all the pieces of the system architecture. There are two artifacts that collect this information together – the system overview diagram and the data dictionary.

The system overview diagram is arguably the single most important artifact of the entire design process, although of course it cannot really be understood fully in isolation. The various descriptors provide the more detailed information that may be required.

The symbols used in the system overview diagram are shown in Figure 7.1.

We start by placing icons for each percept type and action type as well as external data stores, at the boundaries of the figure. If there is a sub-system that is processing data to provide (some of) the percepts prior to them being provided to the agents, then this sub-system can be shown as a box (or multiple boxes), with the percepts originating from within it. Similarly, a non-agent sub-system that manages the details of agent actions can be shown as a box between the agent system entities producing the actions, and the actual actions.

We place within the figure a named agent symbol for each agent type. We then link percepts to agents that use them, and actions to agents that are responsible for them. We also place an incoming link from each agent that writes to an external data store and an outgoing link from each external data store to each agent that directly accesses its data. Double-headed links (arrows at both ends) indicate both read and write.

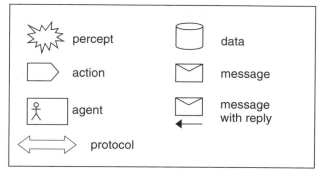

Figure 7.1 Graphical symbols used in the system overview diagram

Once we have arranged the agents and the interface entities, we add shared internal data stores, interaction information between agents and information regarding any additional external sub-systems.

We next place all protocols into the figure, by placing a named protocol link between each pair of agents that communicate via at least one message within a given protocol. If there are additional agent communications that exist outside the specified protocols, we indicate these by placing a named message symbol and linking to and from the relevant agent. If the communication is a message and response pair, then this is shown using the special 'message with reply' symbol.

If external code is being incorporated into the system, this is shown by a box with appropriate message or percept/action interface to some agent(s).

The example system overview diagram in Figure 7.2 depicts the four agents Agent1, Agent2, Agent3 and Agent4. Agent1 initiates ProtocolA involving also Agent3 and Agent2, as well as participating in ProtocolB. It also receives a message from Agent2 and percept P1 from the environment. It reads and writes database DB1. Agent2, as well as participating in ProtocolA and sending a message to Agent1, also interacts with an

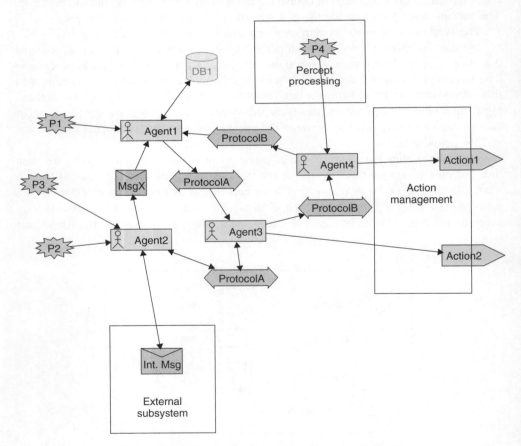

Figure 7.2   Example of system overview diagram

external sub-system via an interface message, and receives two percepts, P2 and P3. Agent3 executes Action2 as well as initiating ProtocolB and participating in ProtocolA. Agent4 receives percept P4 and does Action1, as well as participating in ProtocolB. As only external and shared databases are shown at this level, we see that DB1 is an external database, (read and written by Agent1).

Electronic Bookstore: Case study

The following is a list of percepts, actions and protocols for the bookstore example, followed by the system overview diagram for the example.

Actions: Bank transaction, E-mail stock order, Place delivery request, Request delivery tracking, Send e-mail, WWW page display.

Percepts: Arrival at WWWsite, User input, Bank transaction response, Cheaper price report, Stock arrival, Failed stock arrival, Stock order delay, Regular order trigger, Tracking info, No tracking response, New catalogue.

Protocols: Book finding, Book ordering, Order status querying, Query late books, Stock arrival, Stock delay, Update customer profile.

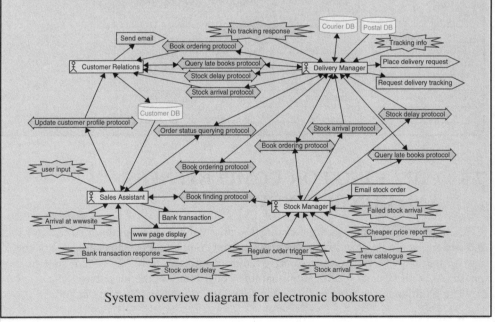

System overview diagram for electronic bookstore

Note that the system overview diagram shows agent *types*, not agent instances.

If desired – and depending on the application – it is possible to provide additional overview diagrams showing agent instances, and partial views of system functionality. Often, these partial views would be focussing on a particular scenario or protocol.

Data Dictionary

The data dictionary is a list of all entities in the design. A data dictionary should be started at the beginning of the project and developed further at each stage.

A data dictionary should be maintained to allow for easy organization and location of the descriptors for the various entities.

One option is to organize the data dictionary into separate sections for agents, capabilities, plans, events and data, organized alphabetically within sections. The other option is to have a flat alphabetical structure. With tool support multiple views (automatically generated) can be provided as well as indexing.

> ☞ TIP: It is useful to develop the data dictionary in such a way that it can be sorted by name, by type, or by location defined. A spreadsheet can be a useful tool here.

Electronic Bookstore: Case study

An excerpt of the data dictionary for the book store is below.

Name	Type	Page	Notes
Bank transaction	action	page 160	
. . .			
Online interaction	functionality	page 141	one per customer
Customer DB	database	page 187	one in system
. . .			
Books DB	database	page 187	one in system
. . .			

7.6   CHECKING FOR COMPLETENESS AND CONSISTENCY

Even more than with the system specification phase, there are a number of ways in which the design should be checked for consistency and completeness. This consistency checking is also likely to incorporate some revision of artifacts from the system specification phase. As with system specification, naming consistency is the most basic check.

The additional checking at this stage falls into a number of main categories:

- consistency between agent descriptors and the descriptors of included functionalities;

- consistency between interaction diagrams, scenarios and protocols;

- consistency of communication specifications: that is, between protocols, system overview and agent descriptors and

- consistency between the descriptors and the system overview diagram.

The first two of the above are essentially vertical checks – ensuring that consistency is maintained with work at an earlier phase. The second two are horizontal checks ensuring consistency between different views within the same phase.

7.6.1 CONSISTENCY BETWEEN AGENTS AND FUNCTIONALITIES

The first check is to ensure that each functionality is assigned to exactly one agent.[1]

Consistency between functionality descriptors and agent descriptors requires a number of basic checks. For the goals and actions fields, the agent descriptor field should be the union of the values in the equivalent field in the functionality descriptors of the included functionalities. Information used and produced by functionalities should be checked against data read, data written and internal data of the agent. This may not be a check that can be automated as apparent inconsistency can result from the fact that during architectural design, early thoughts about data needs have been revised and data representations have developed greater clarity and specificity. However, it is important to check that data needs identified during functionality specification are not inadvertently dropped. Depending on resources and the level of tool support, it may be desirable to update functionality descriptors in the light of further design work. However, as they are not regarded as a final design artifact, necessary for understanding of the final system, this may not be worthwhile.

7.6.2 CONSISTENCY BETWEEN INTERACTION DIAGRAMS, SCENARIOS AND PROTOCOLS

Scenarios, interaction diagrams and protocols are closely related, as is evident from the process described earlier for generating both interaction diagrams and protocols. Things that should be checked are as follows:

- Each scenario should have a corresponding interaction diagram, unless all functionalities involved in the scenario belong to the same agent. It is helpful if naming of interaction diagrams indicates clearly which scenario they are derived from, or if they are tagged in some way to provide this information.

- The interaction diagram corresponding to a given scenario should obey the following rules:

  - For each functionality mentioned in the scenario, the corresponding agent should exist in the interaction diagram.

  - Each agent in the interaction diagram should receive some information (either from another agent or via a percept or trigger) prior to sending out any message. The interaction diagram should really be manually checked to ensure

---

[1] Possibly it is reasonable to have functionalities that are included in multiple agents. However, at the system specification phase, the authors have never actually found this useful. If required, it is a fairly straightforward modification. Of course at detailed design there will very likely be pieces of code – we call them *capabilities* – which will be included or reused in multiple agents.

that an agent has received the relevant information before each interaction – that is, that the information flow makes sense. (This cannot be automated without requiring more detail than is appropriate at this stage of the design.)

- Every interaction diagram that is more than a single message (possibly with response) should be represented within some protocol. This can be checked by mapping each interaction diagram to a scenario and then checking that it is in the *Included scenarios* field of some protocol descriptor. (Single interactions may be shown either by a degenerate protocol specification or by an individual message or message + reply).

- A protocol that is not related to any scenarios should probably result in returning to the system specification phase to specify some scenarios for design completeness. Scenarios are an important part of the final design document for understandability.

- The protocol describing a certain interaction should cover all possibilities. Specifically, if a protocol includes a use case scenario, then it should cover all of the scenario's variations. By 'cover' we mean that the sequence of messages in the interaction diagram corresponding to the use case scenario is one of the sequences that are allowed by the protocol.

7.6.3   CONSISTENCY OF COMMUNICATION SPECIFICATIONS

Communication between agents is specified in three different places within the architectural design. These are within protocols, within the system overview diagram and within agent descriptors. These three specifications need to be consistent – and in some cases, consistency can be actively maintained by a tool during development.

For example, the prototype Prometheus Design Tool ensures that if a protocol defines message passing between two agents, then there is a corresponding link indicated in the system overview diagram. Sometimes, however, it is practical only to automate consistency maintenance in one direction. For instance, it is easy to ensure that a protocol step showing a message between two agents results in a protocol linking these two on the system overview. However, it is less clear that it is desirable to automatically insert a message into a protocol on the basis of the designer indicating a protocol link between two agents. There is insufficient information to completely automate this step – we do not know what message is sent or when in the protocol it should happen – so it may be preferable simply to use the information to direct consistency checking at the end of the phase.

The particular consistencies we want to ensure are as follows:

- If there is a protocol link between two agents on the system overview, then there should be some message between these two agents in the protocol, and the protocol should be mentioned in both agent descriptors. Similarly, if there is a message between two agents in a protocol, it should be reflected in descriptors and the system overview diagram.

- If an agent descriptor lists that the agent *interacts with* another agent type, then that interaction should be evident in some protocol descriptor (or some single message or pair of messages if protocols are not used for this case). Similarly, all agents that are interacted with according to protocol specifications should be listed in the agent descriptor.

7.6.4 CONSISTENCY BETWEEN DESCRIPTORS AND THE SYSTEM OVERVIEW DIAGRAM

There is considerable overlap in the information that is conveyed by the agent descriptors and that conveyed by the system overview diagram. The descriptors contain significant additional information, but for facilitating completeness, they also contain information that can be extracted from other design artifacts. Again, if an automated tool is used, much of this is trivial. If not, it is still straightforward, but surprisingly necessary in order to ensure that the design is indeed consistent. We have already covered above the agent interaction issues that need to be checked between descriptors and system overview. Additional items are as follows:

- Ensure that each agent is represented in the system overview diagram.

- Ensure that all persistent data stores are shown in the system overview and that reads and writes are consistent with agent descriptors, as well as data descriptors.

- Ensure that all percepts are shown in the diagram and are appropriately linked to agents, consistent with agent descriptor declarations.

- Ensure that all actions are shown in the diagram and are appropriately linked to agents, consistent with agent descriptors.

8

Detailed Design: Agents, Capabilities and Processes

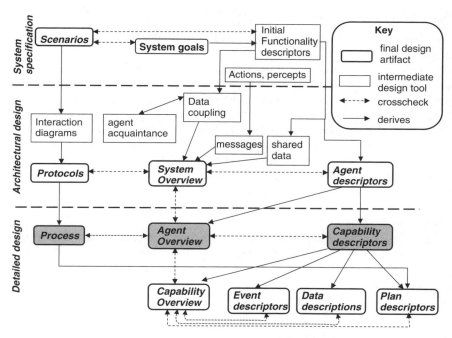

Phases, artifacts and relationships in the design process

During architectural design, we developed an overview of which agents are in the system (the system overview diagram) and how these will interact to achieve system goals (the protocols), as well as descriptors for each agent. In the detailed design, we take these and, for each individual agent, flesh out what capabilities are needed for the agent to

Developing Intelligent Agent Systems   L. Padgham & M. Winikoff
© 2004 John Wiley & Sons, Ltd   ISBN: 0-470-86120-7 (HB)

fulfil its responsibilities as outlined in the functionalities it contains. We also develop the protocol specifications to indicate more of the internal processing of individual agents.

As we get into greater detail (in Chapter 9), we further develop the capability descriptions to specify the individual plans, beliefs and events needed within the capabilities. We also further develop the views that show processing of particular tasks within individual agents. It is during this final phase of detailed design that the methodology becomes specific to agents that use event-triggered plans in order to achieve their tasks. This will be discussed further in the following chapter.

In this initial part of detailed design, the aspects that are addressed are as follows:

1. Decisions regarding the capabilities needed by an agent to fulfil its tasks.

2. Description of the relationships between the capabilities using an *agent overview diagram*.

3. Development of the protocol specifications to indicate some of the internal processing of individual agents.

4. Summarizing information about each capability in capability descriptors.

For the purpose of continuing our running example, we will focus on the Stock Manager agent type.

As with all software engineering, it is appropriate, especially at the detailed design stage, to combine top-down design with bottom-up. In the framework of agent design, top-down design translates to designing the capabilities and their interactions, and then the plans that will be a part of those capabilities. The alternative practice of bottom-up design involves directly designing plans that will accomplish the tasks of the agent, later grouping these into the relevant capabilities. As part of the bottom-up design process, there is also likely to be implementation intermixed with the design. Use of a design tool that generates skeleton code greatly facilitates this process. The reality is that there is usually a mix of bottom-up and top-down development. The larger and more complex the agent is, the greater focus there is likely to be on top-down design, first specifying the capabilities.

In the presentation, we will present first the design of capabilities, then of plans and events. Whichever process is used in the actual development, it is useful in the design documentation to present it from a top-down perspective, to allow the reader to gradually understand greater detail.

8.1 CAPABILITIES

In the architectural design, we have determined what functionalities should be encompassed by each agent, and the list of these functionalities is a natural starting place for the agent capabilities.

Sometimes, there is also functionality akin to 'library routines' that is required in multiple places – either within multiple agents, or within several capabilities within a single agent. Such functionality should also be extracted into a capability that can then be included into other capabilities or agents as required.

Capabilities may be nested within other capabilities and thus this model allows for arbitrarily many layers within the detailed design, in order to achieve an understandable complexity at each level. As the design develops, what was originally a single functionality may well be split into smaller modules, each of which is a capability. It is also sometimes the case that a capability is formed by *merging* two or more functionalities where the functionalities in question are closely related and would result in very simple capabilities if left apart. As previously, the principle of *cohesion* is paramount. Each capability should be a well-defined collection of plans, using particular beliefs or data, which addresses a specific set of goals (or sub-goals) of the agent.

For each capability, we need to determine the goals it is to achieve within the system. If it is derived from a functionality, this information already exists, although there may be some rearrangements at this stage. It is useful, as the design is refined, to also modify the goal and functionality information, if this makes sense.

☞ TIP: **When to modify goals and functionalities developed in System Specification:** This is of course a matter of judgement. If the refinements involve things like splitting a functionality into two capabilities, or adding a new low-level capability that will be included in other capabilities, then there is no need at all to consider modifying functionalities and goals. Indeed, it can be counter-productive to do so, as it can clutter the high-level design with unnecessary detail, making it harder to understand. On the other hand, if the change is significant and reflects rethinking the design, then it can be useful to also consider updating the functionalities. For example, if a goal G is assigned to functionality F but during the design process it is determined that it is better to achieve the goal in capability C (which corresponds to functionality F_2), then it can be useful to update the functionalities by moving G from F to F_2. The main criterion is understandability. If it makes the high level more understandable, then we would move it. If not, then we would leave it, but note at the detailed design the reason for the move.

Electronic Bookstore: Case study

Taking the example of the Stock Manager agent, we develop the capabilities on the basis of the included functionalities. The capabilities and their related goals are as follows:

- **Stock managing** (Goals: Have books in stock, Log books arriving, Log books outgoing, Reorder stock)
- **Pricing** (Goals: Competitive prices, Set prices competitively)
- **Cataloging** (Goals: Update catalogue)
- **Managing competition** (Goals: Competitive prices, Lower book price, Monitor competitive response, Restore book price).

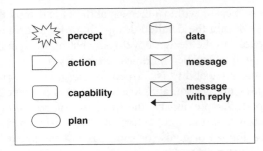

Figure 8.1   Graphical symbols used in the agent overview diagram

8.2   AGENT OVERVIEW DIAGRAMS

The information as to what goal(s) each capability is to achieve, combined with the information of the protocols the agent engages in (from the previous phase), is usually sufficient to enable us to draw the *agent overview diagram*. This diagram shows the relationships between the capabilities, thus providing a top level view of the agent internals. It is very similar in style to the system overview diagram, but instead of agents within a system, it shows capabilities within an agent. This diagram shows the top-level capabilities of the agent and the message flow between these capabilities, as well as data internal to the agent. It is also possible to have some plans at this level if there is some agent functionality that is sufficiently simple that it would be redundant to place it within a capability. Figure 8.1 shows the graphical symbols for the Agent Overview diagram.

To develop the Agent overview diagram, one begins with the interface to the agent as specified in the System Overview diagram. The capabilities are then added to the diagram, and each interface element is linked to at least one capability. The protocol descriptions can then be used to guide the specification of the messages that need to pass between capabilities. Development of initial activity diagrams (based on UML, but described further below) can also assist in this process.

> ☞ TIP: It is often useful to link the names of messages to the name of the sub-goal to which they are related. For instance, when developing protocols, goals in scenarios can guide the naming of messages in the protocols. Similarly, there are typically a number of sub-goals to be achieved, by a particular capability, and it can be useful for the message indicating the need for achievement of the sub-goal to have a name related to that sub-goal. For example, one of the sub-goals of the *Profile Monitor* functionality, within the *Customer Relations* agent, is to *Update customer profile*. When this is needed, the relevant capabilities (*Purchasing, Delivery Manager, Online Interaction*) may well send an *Update customer profile message*. An alternate name for this message could also have been *Request update customer profile*. This helps in linking messages to the goals to be achieved, as specified in the initial phases.

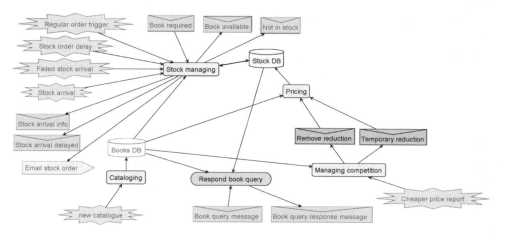

Figure 8.2   Agent overview diagram: stock manager

Electronic Bookstore: Case study

Figure 8.2 shows an agent overview diagram for a Stock Manager agent in the electronic bookstore. Faded entities are the interface to the agent, that is, they depict external entities such as incoming events, actions and percepts.

This agent has the capabilities of Stock managing, Pricing, Cataloging and Managing competition. Note that there is also a single plan for responding to queries on the Books DB and the Stock DB.

8.3   PROCESS SPECIFICATIONS

During system specification, high-level processes were specified by scenarios, which were then refined and more fully specified using interaction diagrams and protocols during architectural design. In detailed design, we also want some mechanism to specify process as well as structure. For this, we use a slight variant of UML activity diagrams. Figure 8.3 illustrates the concepts and the notation of our extended variant of activity diagrams. Note that rather than using swimlanes to separate activities of different agents, as would perhaps be the most obvious modification of UML, we choose instead to focus only on the activity within a single agent, indicating interaction with other agents via the inclusion of messages within the diagram. We believe this is a better choice, partly because it avoids diagrams being overly cluttered, and perhaps more importantly, because it allows for modular development of agents, with shared knowledge only about the interface. It may also be useful in some circumstances to give a process specification diagram which does put together the specifications from all agents involved in that process.

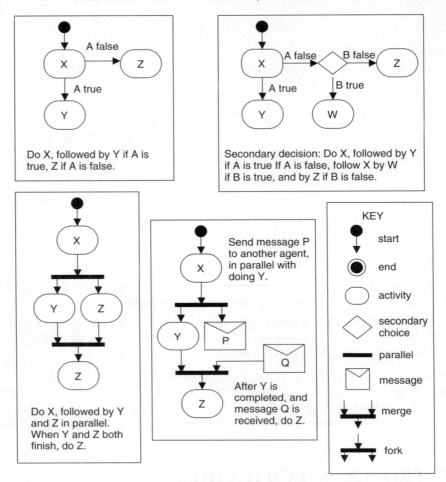

Figure 8.3　Diagram illustrating notation for process specifications

For further information on standard Activity Diagrams, see (Fowler and Kendall 2003).

We identify the process diagrams to be developed by looking at the protocols involving the Stock Manager, as well as the scenarios developed and the goals of the agent. Some process diagrams will not be identified from the protocols as they do not involve significant interaction with another agent.

Electronic Bookstore: Case study

By looking at the scenarios or protocols originating with the Stock Manager agent, and by looking at the goals of the agent, we come up with 5 process diagrams for the Stock Manager agent. They are

- Stock maintenance

- Stock arrival

- Stock delay

- Cataloging

- Pricing reduction.

Stock maintenance and Stock arrival activity diagrams are shown in Figure 8.4.

Notice that stock maintenance is not derived from a protocol. However, the existence of such goals as *Reorder stock* and *Have books in stock* indicate the need for a specification of a process for achieving this. As can be seen, there is no need for a protocol as there is only a single message involved at the end of (one branch of) the activity.

An outgoing message in a process diagram should be able to be mapped either to a process diagram within the receiving agent, or to a possible entry point within a process diagram within the receiving agent. The former is of course simply a special case of the latter, where there is only one entry point and it is the relevant message. Figure 8.5 shows the process for *Updating a user profile* within the Sales Assistant Agent, and the related activity of *Manage customer profile* within the Customer Relations Agent. The *Update customer profile* message from the Sales Assistant activity diagram is one of the possible entry points into the *Manage customer profile* activity diagram in the Customer Relations agent.

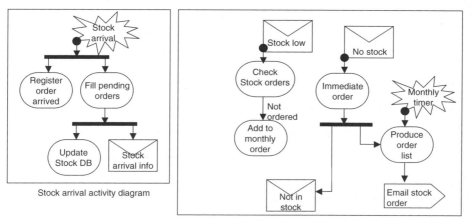

Stock arrival activity diagram

Stock maintenance activity diagram

Figure 8.4   Diagram of some of the processes in the stock manager agent

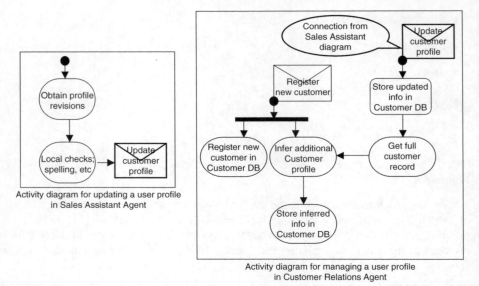

Figure 8.5   Diagram of manage customer profile process in the customer relations agent

8.4 DEVELOP CAPABILITY AND PROCESS DESCRIPTORS

We continue the practice of having structured textual descriptors for each of the significant entities in the system. At this stage, we develop descriptors for each of the Capabilities, and for each of the Process Diagrams. The information for the Capability Descriptor can be extracted almost entirely from existing information, but it is useful to have it gathered in the one place. Use of a support tool can automate most of the process of developing the capability descriptors.

Each Capability Descriptor should have the following fields:

- **Name**:

- **Description**: Brief natural language description.

- **Goals**: Can often be taken from a related functionality, but may need refinement.

- **Processes**: A list of the processes within the agent that the capability is involved in.

- **Protocols**: A list of the system-level protocols that this capability is involved in.

- **Outgoing messages**: A list of messages sent, along with which capability/agent they are sent to, and which goal they relate to.

- **Incoming messages**: A list of messages received, along with which capability/agent they are received from and which goal they relate to.

- **Internal messages**: A list of messages internal to the capability.

- **Percepts**: A list of percepts received directly by this capability.

- **Actions**: A list of actions done directly by this capability.

- **Included capabilities**: Any included capabilities, along with the goals associated with those capabilities.

- **Data used: Imported**: Data imported to the capability

- **Data produced: Exported**: Data exported from the capability.

- **Data internal**: Data used internally.

- **Included capabilities**: Capabilities within this capability.

- **Included plans**: Plans within this capability.

- **Notes**: Any relevant design notes.

Electronic Bookstore: Case study

Here is the capability descriptor for the Ordering capability within the Stock Manager Agent.

Name:

Description: This capability manages the ordering of stock from suppliers – either on a regular basis, or if necessary when stock runs out

Goals: Order stock

Processes: Stock maintenance

Protocols: Book ordering protocol

Outgoing messages: Book available, not in stock

Incoming messages: Book required

Internal messages: Modify monthly order, decide supplier, Get number required, No stock, stock low

Percepts: Regular order trigger

Actions: E-mail stock order

Included capabilities: None

Data used: Imported Pending Orders, Books DB

Data produced: Exported Stock Orders, Pending Orders

Data internal: Monthly Order

Included capabilities: none

Included plans: Check stock, Add to order, Out of stock response, Number by index, Get number by price, Get number by sales, Decide supplier by time, Decide supplier by price, Delete items, Add to supplier order, Build monthly orders

Notes: (none).

The Process Descriptor also contains information that is available in the rest of the system, but can be useful to have accumulated in one place, so that it can be viewed along with the Process Diagram to obtain additional information such as the receiving agents of a message.

We use a Process Descriptor with the following fields:

- **Name**:

- **Description**:

- **Activities**: May be existing (sub-)goals, or newly identified such

- **Triggers**: List of goals/messages/percepts

- **Messages**: <message name, to-agent>

- **Protocols**: Those protocols this process participates in

- **Notes**: (none).

Electronic Bookstore: Case study

Here is the process descriptor for the 'Stock maintenance' process, within the Stock Manager Agent.

- **Name**: Stock maintenance.
- **Description**: The activity whereby there is an attempt to maintain sufficient stock to immediately fill orders. The activity is responsive to immediate demands as well as maintaining stock levels from month to month.
- **Triggers**: Book required, Monthly timer.
- **Activities**: Check Stock DB, Immediate order, Add to monthly order, Produce order list.
- **Messages**: <Update customer profile, to Customer Relations Agent>, <Book purchase, to Delivery Manager Agent>
- **Protocols**: Book ordering protocol, Stock arrival protocol, Query late books protocol.
- **Capabilities**: Stock managing.
- **Notes:** (none).

9

Detailed Design: Capabilities, Plans and Events

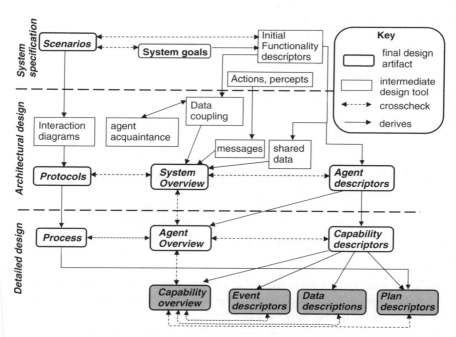

Phases, artifacts and relationships in the design process

At this final stage of the detailed design, each capability is broken down either into further capabilities or, eventually, into the set of plans that provide the details of how to react to situations, or achieve goals. At this stage, a number of details regarding the implementation platform become important. We focus on Belief-Desire-Intention (BDI)

Developing Intelligent Agent Systems   L. Padgham & M. Winikoff
© 2004 John Wiley & Sons, Ltd   ISBN: 0-470-86120-7 (HB)

platforms in particular, which are characterized by a representation that has hierarchical plans with triggers, and a description for each plan that indicates the context in which it is applicable. BDI systems choose among the plans that are applicable, backtracking to try another plan if the one initially chosen does not succeed.

Although a number of the issues we address at this point are specific to the way BDI systems work and are written, the principles are easily adapted to a range of other agent systems. In particular, any system using agents with plans that react to events can easily be accommodated. At the same time, if the precise development platform is known, then additional issues can be considered and noted during design.

The details we provide here are oriented towards BDI systems, but not towards any particular BDI implementation platform. We do, however, provide examples of additional detail that could be added for a particular platform, using JACK Intelligent Agents as an example platform.

The main steps covered in this final stage are as follows:

1. Further decomposition using capability overview diagrams

2. Sub-tasks and alternative plans
   - Identifying context conditions
   - Coverage and overlap

3. Developing the events and messages

4. Action and percept details

5. Details of data.

9.1 CAPABILITY OVERVIEW DIAGRAMS

In the previous chapter, we developed agent overview diagrams. A further level of detail is provided by capability diagrams that take a single capability and describe its internals. At the bottom level, these will contain plans, with *internal messages* providing the connections between plans, just as they do between capabilities. At intermediate levels, they may contain nested capabilities or a mixture of capabilities and plans. These diagrams are similar in style to the system overview and agent overview diagram, although one of the incoming messages to a plan needs to be identified as the triggering message. Each plan must have exactly one triggering message.

Figure 9.1 shows a capability diagram for the *Stock managing* capability of the Stock Manager agent shown in the previous chapter (Figure 8.2 on page 103). This capability was quite large, as can be seen from the number of incoming and outgoing items on the Stock Manager agent overview diagram from the previous chapter. Consequently, we choose to break it down into the three further capabilities, *Ordering, Delay handling* and *Handling new stock.*

Figure 9.2 shows the plans that realize the *Ordering* capability. Consistent with the parent diagram of the *Stock managing* capability, it has one incoming message, two outgoing messages, an action and four external data sources. In addition, it introduces one new data source internal to this capability – the *Monthly Order.*

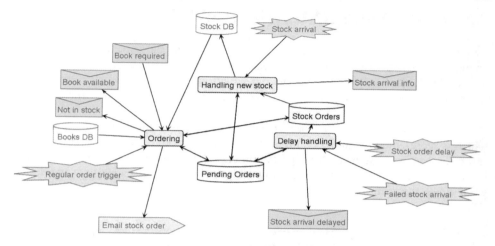

Figure 9.1   Capability diagram: Stock managing capability

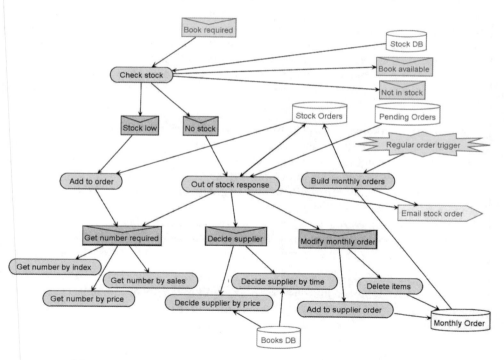

Figure 9.2   Capability overview diagram: Ordering capability

9.2   SUB-TASKS AND ALTERNATIVE PLANS

At the lowest level of detail, each incoming message to the capability must have one or more plans that respond to that message. Multiple plans responding to the message provide multiple ways of reacting to the situation. This allows each individual plan to be simple and straightforward, responding to the relevant message for a particular situation.

Each plan can typically be broken down into some number of sub-tasks, each of which is represented by an internal message. Depending on the implementation platform, sub-tasks can be combined using programming control structures – sequence, parallel, if, or, and so on. These control structures are not shown in the capability overview diagram – like the other overview diagrams, it captures only static structure.

Electronic Bookstore: Case study

In Figure 9.2 of the *ordering* capability, we see that the *Check stock* plan has four messages or sub-tasks: *Stock low, No stock, Book available* and *Not in stock*. These actually represent two cases requiring ordering of stock, as well as a third case not requiring stock ordering. In one case there is no stock, with an internal message generated to handle this situation and a message sent to the entity requiring the book, stating that it is not in stock. In the second case, stock is low, and so a message is sent back saying the book is available as well as an internal message being generated to handle the low stock situation. The third case (not requiring any ordering of stock) simply sends the *Book available* message.

The two stock-ordering situations trigger two separate plans to deal with the two situations. Figure 9.3 shows the items relevant to the situation in which stock is low. In this situation, the *Stock low* message triggers a plan *Add to order*, which adds the book to the monthly order form, after first checking in the *Stock orders* DB that it is not already ordered and awaiting delivery, and the *Monthly Order* DB to ensure it is not already on this month's order list. *Add to order* then has three sub-tasks: *Get number required*, which determines the number of books to order, *Decide supplier*, which determines which supplier to order from and *Modify monthly order*, which adds the required number of books to the monthly order for the relevant supplier.

The *Get number required* message is an example of a situation in which there are several different plans or ways of achieving this sub-task. *Get number by index* simply obtains the number on the basis of a categorization of the book, and a standard number of copies to order for this book category. *Get number by price* determines the number of copies to order on the basis of the price of the book – perhaps not wishing to have too much money tied up in many copies of expensive books. *Get number by sales* looks at the sales history for the book and determines an order number on the basis of this information. The availability of these different plans, allows the system to be flexible and choose depending on circumstances. For example, if

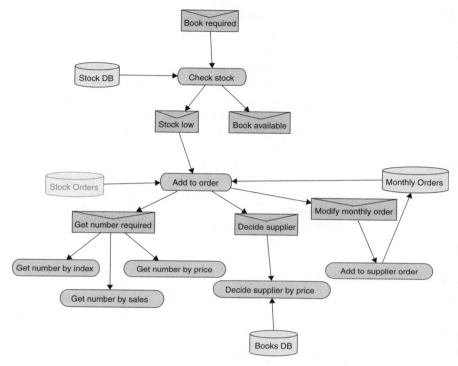

Figure 9.3   Low stock situation within the Ordering capability

there is a cash flow problem, it may be most appropriate to use *Get number by price*. This decision can be made at run time. (We note that each of these plans would actually need access to data that is not currently shown in the diagram).

It is often the case that initially only one plan may be provided for achieving a particular sub-goal. Usually, this is the most straightforward or standard way of doing something. Later, additional plans can be added to address more unusual situations, or to provide alternative approaches that can be useful in some situations.

9.2.1   IDENTIFYING CONTEXT CONDITIONS

Each plan is triggered by a specific *event*. This event may be the arrival of a percept, arrival of a message from another agent, or an internal message or sub-task within the agent. We use the term event to include each of these, as it is also a commonly used term in many agent systems. If there are several plans that could be triggered by a given event, then it is important to specify the conditions or situation under which the various plans are applicable. We call this the *context condition*, and in many agent systems (and all BDI systems), this maps quite directly to an implementation concept.

APPLICABILITY

The context condition most often represents information about the state of the environment, which makes the plan suitable for use in that situation. For example, a plan to recompense a customer for a lost book by immediately sending a new book may have as a context condition that the required book is in stock. An alternative plan may provide the customer with a refund and have as a context condition that the book is not in stock. Additional plans providing more options may be added later.

Information required in context conditions has implications for other parts of the design, in particular, data representations. Exactly what is possible will depend to some extent on the implementation platform. Some systems provide a specialized data structure for representing agent beliefs, which can then be accessed in context conditions. There may also be internal data structures in the host programming language, which contain information about the environment. There may also be functions that can be run as needed as part of a check of context condition. Care should be taken to specify context conditions precisely at this stage and to ensure that the design includes the relevant data.

MULTIPLE BINDINGS

Many BDI systems provide the ability to bind variables as part of a context condition, giving the possibility of multiple plan instances, with different bindings of some variables. For example, a plan to courier a book requires that the destination be in an area that is served by a courier in the database. There may well be several such couriers. Binding of a courier variable to these different options gives multiple plan instances, each using a different courier. This gives flexibility and robustness, in that if a plan to use one courier fails, because perhaps they do not respond, then another plan (with a different courier) will be tried.

FULL SPECIFICATION

In some cases, a context condition should be specified even if there is only one plan available. If there is some situation that is necessary for the successful execution of the plan, then this should be specified as a context condition, even if there is no other option available. (By doing this, the system is able to efficiently and effectively recognize a failure situation, and possibly remedy it by making other choices elsewhere.)

Sometimes, inexperienced agent programmers fail to represent context conditions that they expect to be true, on the basis of previous actions. However, it is important to remember that an agent does not have complete control of the environment, and external events can change the environment during the time an agent is pursuing a series of plans towards a particular goal. Representation of necessary context conditions helps ensure the correct functioning of the agent.

Electronic Bookstore: Case study

Returning to the example of the different plans for *Get number required* in our example in Figure 9.2, we could define context conditions as follows:

Get number by index
Context: Book has category assigned
Data requirements: Category field in Stock DB; Data structure (*Ordering-categories*) mapping categories to a default number to order.

Get number by price
Context: Wholesale book price greater than $100
Data requirements: Wholesale book price as a sub-field of supplier in Books DB.

Get number by sales
Context: Sales for month greater than default number orders for category
Data requirements: Data structure (Monthly sales) that holds number of sales per book; structures for determining default (index) order values as for *Get number by index* plan.

9.2.2 COVERAGE AND OVERLAP

Having multiple plans to respond to an event is a powerful mechanism that facilitates robustness, flexibility and modularity. However, it does require that the designer consider potential interactions between the plans. *Coverage* is the term we use to refer to the concept of whether, for a given event, there will always be some plan with a matching context condition. *Overlap* refers to whether it is possible, in some situations, to have more than one plan that is applicable, necessitating a choice between them.

If there is a plan with no context condition, then coverage is guaranteed. It can be useful to have such a plan to ensure coverage. However, it is important, as noted previously, to not ignore (and fail to represent) conditions that are actually required for successful execution, but which one expects will hold. Coverage can also be guaranteed if there is a pair of plans whose context conditions are the complement of each other. For example, if one plan has a context condition that book price is less than $100, and another has a context condition that book price is greater than or equal to $100, then there is coverage. If, however, the second plan has only book price greater than $100, then (if these are the only two plans) there is not (complete) coverage, as in the case where book price is exactly $100, neither plan will be applicable. It is important for the designer to do a careful analysis of coverage to ensure that, unless intended otherwise, there is no gap in coverage. If it is intended that there be no response in some situations, then a gap in coverage may be appropriate. However, it is important to check how this is handled by the implementation platform and to implement accordingly; in some systems, it may be necessary to have an 'empty' plan that does nothing, to obtain the desired behaviour.

A related concept to coverage is that of *overlap*. We say that there is overlap in response to a trigger if it is possible for there to be multiple plans that are applicable in some situation. There are two mechanisms by which this can happen. One is when two context conditions overlap, that is, there is some situation in which both could be true simultaneously. The other is where the context condition contains variables that can potentially take on multiple values at a single point in time, leading to multiple plan instances of a single plan type. Overlap is a kind of redundancy and can be very useful in building a system that is robust and able to recover from failures. However, it is very important that it be considered and understood at design time.

> ☞ TIP: Interactions between context conditions that have not been foreseen are one of the very common bugs that arise in agent systems. Sometimes there is a failure or lack of response, due to an unintended gap in coverage. At other times, there is an unintended overlap in context conditions, leading to a different plan being executed than that which was intended. Careful and systematic analysis during detailed design helps eliminate these bugs.

Preferences

If there is overlap, one plan will be chosen (initially) from all those that are applicable. It may be that although several plans are viable, there are preferences as to which should be tried first. The mechanisms available for specifying and controlling this vary from system to system. However, it is important that the designer consider whether such preferences exist, and if so, what they are. If the implementation platform lacks a mechanism for controlling the order in which plans from an applicable set are chosen, then the designer must weigh up the advantages of controlling ordering by tighter specification of context against the advantages of robustness by not excluding less preferred but possible plans.

9.3 EVENTS AND MESSAGES

As indicated, we use the term *event* to denote things that can trigger the choice and execution of a plan. These can be the messages that we have identified between agents, the percepts coming from the environment, or the messages internal to an agent, indicating sub-tasks or messages between plans and/or capabilities.

The most important issue associated with the events is the precise specification of the information carried by that event. Again, the exact way in which this is carried is dependent on the implementation platform. The important design issue is to specify clearly what information is required as part of the event. Both the type of information, and, where relevant, the allowable values should be specified. In some cases (for percepts or messages between agents), this will have been specified at an earlier stage. However, it may need to be refined or added to during detailed design. In addition, it is important at this stage to add additional information such as that regarding coverage and overlap. Many new events (or internal messages) are developed as the plans take shape and sub-tasks are identified.

For example, if there is a book order event, presumably it is important for that event to carry the information as to which book is being ordered. It may also be necessary to carry information about who is ordering and the delivery address. This information has possibly been identified during architectural design. As a plan is being developed, it becomes evident precisely which information is required, and as a result, the information to be carried by the event is refined. Some events may only serve as a trigger and may not need to carry additional information.

For messages between agents, it is also important to specify whether a reply is expected and the details of that reply.

Events representing percepts from the environment must specify the information carried as part of the percept. This may have been extracted from raw data as part of a processing phase.

Electronic Bookstore: Case study

Referring to Figure 9.3, we see six different events. Many of these events will need to carry the information regarding the book under consideration. The exception is the *Book available* message, which is a response to a request, and therefore may not need to explicitly carry this information.

The *Modify monthly order* event would need, in addition to book id, to carry the following data:

- supplier

- number of copies.

9.4 ACTION AND PERCEPT DETAILED DESIGN

Actions and percepts have been identified during earlier stages. If there is processing of incoming data required to obtain percepts, this needs to be designed and specified. If a non-agent approach is being used to process the data (as is common in, say, image processing), then design can be done using a notation and methodology suitable for the approach being used. What is important is that the design of this part of the system is not forgotten, and, in particular, that wrong assumptions are not made as to what will/can be provided.

Like percepts, actions may be simple or complex. They may be simple system calls that invoke well-known and supported functionality, such as sending an e-mail or, alternatively, they may require customized design and implementation. Particularly in physical systems, actions often require significant work in order to obtain satisfactory behaviour. For example, in robocup, a move-to-point action required addressing issues such as how close to the point would be considered 'at' the point, what velocities and accelerations should be used, whether these should be parameters to the action and if so how they should be dealt with, as well as issues to do with the feedback loop between effectors and code. If actions are complex, then design documentation regarding them is an important part of the design of the system as a whole. Like percept processing, the notation and methodology used for designing the action components should be appropriate for the approach being used.

9.5   DATA

Data representation is an important part of any system design activity, and agent systems are no exception. However, as mentioned earlier, we do not focus on this aspect as existing techniques are suitable for this purpose. Nevertheless, data representations must be decided and specified, using some appropriate methodology.

Some agent development environments do offer some specialized support for data representation, such as the *beliefsets* available in JACK. If available, such specialized data representations can be useful as they typically provide some additional functionality. However, this is again an implementation platform issue. For arbitrary data structures, object-oriented class diagrams and access methods can be an appropriate representation.

What is important at the detailed design phase is to ensure that all significant data structures are well specified. All data required by plans should be identified and the location of that data should be specified.

9.6   DEVELOP AND REFINE DESCRIPTORS

The final design artifacts required are the individual descriptors that provide the detailed information for each item. The new descriptor type introduced during detailed design is the plan descriptor. Other descriptor types often obtain additional fields at this stage. For example, agents and capabilities need fields added to indicate which plans are included. Messages and percepts need to be extended with information about such things as coverage and overlap. In addition, many new instances will be added at this stage, in particular, many new internal messages representing sub-tasks, and many new data descriptors.

The descriptors provide the details necessary to move into implementation. Some of these details will depend on aspects of the implementation platform, while others are more generic. For example, if the context in which a plan type is to be used is split into two separate checks within the system being used (as is the case in JACK), then it is appropriate to specify these separately in the descriptor. Fields added to percepts and messages, specifying what information an event carries will depend on the structure of events within the system, and so on. The guiding principle is that these descriptors should (a) carry enough information for the reader to thoroughly understand the capability overview diagram, and (b) carry sufficient information to move directly to implementation.

The *plan descriptors* we use provide an *identifier*, the *triggering event type*, a *context* specification indicating when this plan should be used, the *plan steps* as well as a short natural language *description*, *messages* received and sent and a list of *data* used and produced. At this stage, data lists should always be specifically tied to particular data descriptors, rather than being generic descriptions.

The *coverage and overlap* information discussed earlier should be added to all message and percept descriptors. An event is *covered* if there is always at least one handling plan that is applicable; that is, for any situation, at least one of the matching plans will have a true context condition. There is *overlap* if it is possible to have multiple plan instances available to respond to this event. No overlap means that there is always at *most* one plan that is applicable in any situation. If there is overlap, then some notes should be made regarding this, and the preferences between overlapping plans should be made clear, if there are any.

If the implementation system allows sub-typing of events, (as can, for example, be done using JACK's *relevance* field), then this should also be added as additional information in the message/percept descriptor.

Data descriptors should specify the fields and methods of any classes used for data storage within the system. If specialized data structures are provided for maintaining beliefs, these should also be specified. The data dictionary should also be updated and checked at this stage.

Electronic Bookstore: Case study

Following is an example of a plan descriptor for a plan to add a book to the stock order.

Plan: Add to order
Description: Order more stock of a book.
Trigger: Stock low
Context: Book not already ordered
Data used & produced: Reads Stock Orders and Monthly Order
Goal: Have books in stock
Failure: never fails
Failure recovery: not applicable
Procedure:
1. Determine how much stock to order (Get number required)
2. Determine which supplier to use (Decide supplier)
3. Add the order to the monthly order (Modify monthly order).

9.7 CHECKING FOR COMPLETENESS AND CONSISTENCY

As with *system specification* and *architectural design*, the *detailed design* also needs to be checked for completeness and consistency. As with previous phases, this is greatly assisted by the use of a tool that can automate a large part of this work. As with earlier phases, naming consistency is important, but we do not further discuss it here. At this stage, there is a proliferation of entities and many will have related names, so it is critical to know that only the exact same name refers to the same entity.

We structure the things that we check at this stage into four main areas:

1. Checking the completeness of each agent:
   Does it implement the required functionality? Are all messages required for its specified participation in protocols generated and received?

2. Nothing is missing or redundant:
   Ensure that all data is both used and produced (except for external data, where it is possible for it to be only used or produced). Ensure that all messages are both sent and received.

3. Checking the consistency between final design artifacts:
   We need to ensure consistency of interfaces between levels. We require consistency between overview diagrams and descriptors, and we check consistency of plan details.

4. Following through important scenarios

As mentioned previously, these checks are a combination of horizontal checks, within the artifacts generated during this phase, and vertical checks with artifacts generated during previous phases.

9.7.1   AGENT COMPLETENESS

One of the most basic checks is to ensure that each agent is in fact covering the full range of functionality that was assigned to it in the architectural design. Often, each functionality is directly represented by a capability, in which case it is straightforward to ensure that no functionalities have been missed.

It is also a good idea to ensure that each message required by the protocol specifications in which the agent participates is in fact received or generated by the agent.

Plans implement the details of the functionality provided and therefore it is wise to carefully review the subsets of plans that implement particular processes. In particular, it is useful to review the set of plans that respond to a single trigger, to ensure that all situations are covered by some plan (or that it is acceptable if this is not the case). Overlap between plans can also be reviewed at this stage.

9.7.2   MISSING OR REDUNDANT ITEMS

As a system is developed, it is extremely common to end up with some redundant or missing information. As functionality is trimmed, or different ways are decided on for doing things, items that were previously needed are no longer relevant. As functionality is added, there is addition of items, but sometimes not all of them are fully integrated.

Missing information will of course show up in some way in the implemented system, possibly manifesting as some kind of an error. Redundancy, however, may never be noticed – resulting in additional work for no benefit during implementation, and code that is harder to maintain.

The specific items that we check in this area are messages, data, actions and percepts.

MESSAGES MUST BE BOTH SENT AND RECEIVED

All messages in the system must be ultimately sent from at least one plan. Similarly, all messages must be received by at least one plan. Usually this will be as a plan trigger, although some messages may be received within a plan body. This check can be done by looking at plan descriptors and their incoming and outgoing messages. If an incoming message is not also a trigger, then the procedure should be checked to ensure there is a mechanism for receiving the message.

DATA MUST BE BOTH USED AND PRODUCED (USUALLY)

Usually, it is the case that if data is produced, we would want to see that it is also used. An exception may be if the data is being produced purely for external purposes, for example, a summary report of the number of storms in a year from a weather bureau, or a statistical report on sales from our electronic bookstore.

Certainly, data that is used must be produced or obtained somewhere. Data can come basically from three sources. It can be produced by some entity within the system; it can come from environmental information via percepts; or it can come from external data stores. If it is not being produced within the system, it is a good idea to check carefully the source.

ACTIONS AND PERCEPTS

It is of no use having actions defined (and perhaps implemented) that are never actually executed. Similarly, spending time inputting percepts, and perhaps collecting and processing data for these, is wasted effort if they are not used in some way.

During architectural design, it was established which agents were receiving which percepts and executing which actions. Now it is important to ensure that some plan within the relevant agent(s) actually executes each action. For each percept, it is important to ensure that within the relevant agent(s) it either triggers some plan (which is the most common case) or that it automatically updates some knowledge or belief store, without use of a plan, or that it is accessed from within a plan body. Checking off percepts that are used as plan triggers is easily automated via descriptors and/or overview diagrams. The remaining cases should be checked manually.

9.7.3   CONSISTENCY BETWEEN ARTIFACTS

As the design is developed, it is important to continue to monitor consistency between the interfaces at the more abstract and the more detailed levels, as well as consistency between different kinds of representations.

CONSISTENCY BETWEEN OVERVIEW DIAGRAMS

The interfaces between the various overview diagrams can be simply and effectively checked for consistency. All incoming and outgoing links from an agent in the system overview diagram must translate to the same incoming and outgoing links within the

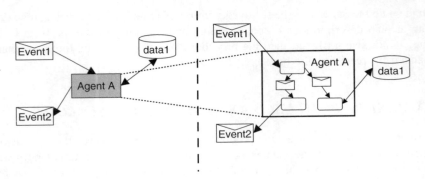

Figure 9.4   Interfaces should match for every entity at different diagram levels

agent overview diagram. Similarly, the incoming and outgoing links from a capability within an agent overview diagram must be identical to the incoming and outgoing links in the capability overview diagram for that capability. The same principle applies to nested capabilities. Figure 9.4 illustrates this principle: the events and data that form the agents interface in the system overview diagram (left) also appear in the agent overview diagram (right).

This check sounds very trivial, but it is surprising how often an interface item is inadvertently added or dropped.

> ☞ TIP: The authors have found that large numbers of minor errors and omissions that would most likely percolate into code are easily found by mechanical and routine consistency checking, much of which can be automated. We have also found that design documents with widespread inconsistencies are extremely difficult to understand if one has not been involved in the design process.

CONSISTENCY BETWEEN OVERVIEW DIAGRAMS AND DESCRIPTORS

Consistency between all overview diagrams and the relevant descriptors should be performed, in a manner similar to that described in Section 7.6.4. Capability descriptors should be checked against the agent overview diagram, or in the case of nested capabilities, against the relevant capability overview diagram. If a capability is used in more than one place, then it should be checked against all relevant overview diagrams.

Plan descriptors should be checked against the capability overview diagram (or in some cases, the agent overview) to ensure that incoming and outgoing messages and data uses are consistent. With regard to plan descriptors, it is also useful to check some aspects against the procedure description. For example, it should be evident within the procedure description where each outgoing message is posted. Similarly, for any incoming message that is not the trigger, it should be evident where it is accessed. A similar check can be made with respect to actions and percepts if relevant.

Descriptors for messages, data stores, percepts and actions should also be checked for consistency with overview diagrams.

CONSISTENCY OF SCENARIOS, PROCESSES AND PLAN SETS

Scenarios were developed at an early stage to capture important aspects of the system functionality. Before detailed design can be considered complete, a check should be made to ensure that these scenarios are captured by processes and plan sets within each of the relevant agents. It is not uncommon that in the process of detailed design, scenarios do change somewhat. Depending on the project, it can be important to update the scenarios, as they are important in helping to understand a design. If they are not updated, they should at least be marked as no longer current when inconsistencies are detected, and it is determined that it is the scenario that should be changed.

PROCESSES CONSISTENT WITH PROTOCOLS AND OVERVIEW DIAGRAMS

Processes are a detailed description of the dynamics of various activities within a particular agent. They should be checked especially against the protocol related to the activity, particularly to ensure that messages in and out are consistent with the message exchanges indicated in the relevant protocol. Similarly, triggers for the activity should be found (as percepts or messages) on the agent overview diagram, unless they are confirmed to originate within the agent.

These checks are not readily automatable, but structured manual checking is quite effective.

9.7.4 IMPORTANT SCENARIOS

It can be useful to do a 'design walk through' of key scenarios together with a colleague. Flaws can often be seen at this stage and it is preferable to rethink and remedy them within the design.

10

Implementing Agent Systems

The design process of Prometheus results in descriptors and diagrams that describe a range of design entities such as goals, functionalities, agents, capabilities, plans, percepts, actions, messages, data, protocols and scenarios. However, not all of these design entities are carried through to implementation. For example, functionalities are used to determine the agent types but they do not correspond to any run-time entity. Roughly speaking, the entities that are implemented are those that are produced in the detailed design phase: namely, agents, capabilities, plans, messages and beliefs; as well as actions, percepts and goals. In this chapter, we briefly look at how we undergo a transition from detailed design to implementation and how certain agent platforms support a very direct mapping of these concepts.

Clearly, the concepts of messages, plans and beliefs (as well as agents and capabilities) do not map directly into object-oriented languages such as C++ or Java. Although it is possible to implement an agent-oriented design using an object-oriented language, this is akin to trying to realize an object-oriented design in a non-OO language such as C: awkward and difficult to maintain.

10.1 AGENT PLATFORMS

Fortunately, there exist a range of 'agent-oriented' programming languages. Generally, these languages also provide libraries and infrastructure, such as naming services ('white pages'), and so we shall use the term 'agent platform'. When considering which agent platform to use, it is important to be aware that not all agent platforms support plan-based agents. Roughly speaking, there are three classes of agent platforms:

1. Those that focus on internal agent reasoning and support plans, goals, and so on.

Developing Intelligent Agent Systems   L. Padgham & M. Winikoff
© 2004 John Wiley & Sons, Ltd   ISBN: 0-470-86120-7 (HB)

Examples: PRS[1] (Georgeff and Lansky 1986; Ingrand et al. 1992), UMPRS[2] (Lee et al. 1994), JAM[3] (Huber 1999), JACK[4] (Busetta et al. 1999), DECAF[5], Zeus[6], AgentBuilder[7] and JADEX[8] (which is an extension of JADE[9]).

2. Those that focus on inter-agent communications. These usually provide infrastructure for inter-agent communication (for example, a way of sending messages in a certain syntax) as well as facilities for locating agents on the basis of names (white pages) and/or a description of the service that the agents provide (yellow pages). More recent platforms in this class tend to conform to the FIPA[10] standards. Examples: JADE[11], Zeus[12], OAA[13] (Cheyer and Martin 2001)

3. Those that focus on mobile agents.
   Examples: Grasshopper[14], D'Agents[15], Aglets[16]

Although inter-agent communication is a specialty of class 2 platforms, all (realistic) platforms provide some support for agent communication (for example, although JACK is listed as being a class 1 platform, it provides support for agent messaging and for locating agents by name). A few platforms provide support for multiple areas. For example, a FIPA-compliance module for JACK is available[17]. Another example is JADEX[18], which extends the FIPA-platform JADE with plans, events, goals and beliefs. Another example is JAM, which is primarily a Belief-Desire-Intention (BDI) architecture (class 1), but provides support for agent mobility.

The first class is most useful in terms of providing support for implementing designs developed by following the full Prometheus methodology (including final detailed design).

The aim of this chapter is not to provide a detailed survey of a wide range of platforms. The range and status of available agent platforms changes and the most up-to-date information can be found online. Thus, we give pointers to existing surveys rather than provide a survey that will be out of date by the time this book is published. Two sites that provide extensive lists of tools are

---

[1] *http://www.ai.sri.com/~prs/*
[2] *http://www.marcush.net/IRS/irs_downloads.html*
[3] *http://www.marcush.net/IRS/irs_downloads.html*
[4] *http://www.agent-software.com*
[5] *http://www.eecis.udel.edu/~decaf/*
[6] *http://more.btexact.com/projects/agents/zeus/*
[7] *http://www.agentbuilder.com/*
[8] See *http://sourceforge.net/projects/jadex*
[9] *http://sharon.cselt.it/projects/jade/*
[10] Foundation for Intelligent Physical Agents, *http://www.fipa.org*
[11] *http://sharon.cselt.it/projects/jade/*
[12] *http://more.btexact.com/projects/agents/zeus/*
[13] *http://www.ai.sri.com/~oaa/*
[14] *http://www.grasshopper.de/*
[15] *http://agent.cs.dartmouth.edu/*
[16] *http://www.trl.ibm.com/aglets/*
[17] From *http://www.cs.rmit.edu.au/agents/protocols/*
[18] See *http://sourceforge.net/projects/jadex*

EXAMPLE 127

- *http://www.cems.uwe.ac.uk/~rsmith/ECOMAS/agent_toolkit_list_(courtesy_of_bt). htm*[19]

- *http://www.agentbuilder.com/AgentTools/index.html*

Additionally, Luck *et al.* (2004) discusses a number of platforms.

In the remainder of this chapter, we focus on illustrating how the detailed design that has been produced can be implemented using a plan-based agent system. Specifically, we show the close match between the results of the detailed design and the concepts supported by agent platforms such as JACK[20]. We begin by briefly describing the JACK platform, then discuss issues in mapping the detailed design entities produced by following Prometheus to JACK and give JACK code corresponding to a small part of the detailed design for the book store example.

10.2 JACK

JACK is an agent platform based on the BDI model. It views an agent as having plans that are triggered by events where messages are viewed as a specific sub-type of event. A JACK program consists of declarations of entities: agents, capabilities, plans, events and beliefsets (also termed 'databases'). Each declaration links to other entities; for example, an agent declaration specifies what plans and capabilities it contains, what beliefsets it has, what messages it receives and sends and what events it posts internally.

Many of the Prometheus concepts map directly to JACK. For example, a Prometheus agent just becomes a JACK agent. JACK also supports capabilities (Busetta *et al.* 1999) and so capabilities in Prometheus map directly to JACK capabilities.

JACK does not have concepts corresponding to percepts and actions. We represent percepts as events. Actions are simply performed within the plan body using Java code.

Also, BDI systems (including JACK) do not implement goals directly. Instead, they model the acquisition of a new goal as an event. Thus, goals are realized by creating an event type corresponding to the goal. As discussed by Thangarajah *et al.* (2002), this has problems such as top-level goals being dropped if there are no applicable plans in the current situation.

10.3 EXAMPLE

As an example, we focus on the highlighted portion of Figure 10.1[21], which sits within the Ordering capability that is within the *Stock managing* capability within the Stock Manager agent type (see Figure 10.2). This excerpt shows an (internal) message (*Stock low*) that triggers a plan (*Add to order*). The plan makes use of the *Stock Orders* and *Monthly Order* databases, and posts a number of internal messages that trigger further

---

[19]The text '(courtesy_of_bt)' is part of the URL.

[20]A commercial product of Agent Oriented Software, *http://www.agent-software.com*

[21]The figure focuses on the *Add to order* plan and its context and is incomplete – the complete capability overview diagram is in Figure 10.5 on page 132.

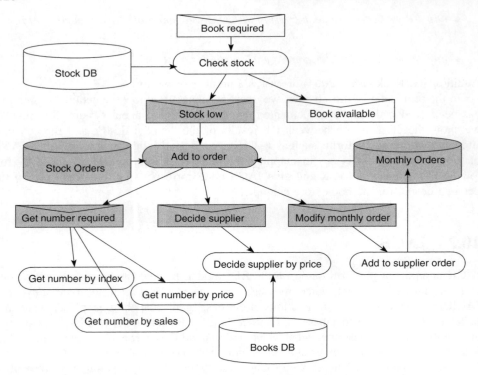

Figure 10.1   Low stock situation within the Ordering capability

plans. The *Stock Orders* database is internal to the agent (but external to the capabilities). The *Monthly Order* database is internal to the *Ordering* capability.

In addition to the database, there is also an object type *book record* that is carried by the messages. This contains information about the book that has low stock, including the book identifier (*bookid*), the *wholesaleprice* and the (optional) *category* of the book.

We consider in turn agents, capabilities, data, messages/events[22], and plans. For each entity type, we give an example descriptor, discuss the issues in mapping the design entity to JACK code and give sample JACK code. The code we give is skeleton code and is intended to illustrate how individual design entities map to JACK code, not to be complete and detailed[23].

10.3.1   AGENTS

We begin with declarations for the agent and two capabilities shown in Figure 10.2. The plans, data and events are all within the *Ordering* capability that in turn is within a *Stock managing* capability that is within a *Stock Manager* agent.

---

[22]Prometheus uses the term 'message', whereas JACK uses the term 'event'.

[23]JACK does not allow for spaces in names and so we either remove spaces (and use capitals to indicate multiple words) or replace spaces with underscores.

EXAMPLE 129

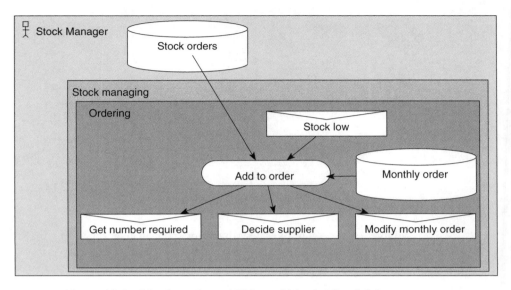

Figure 10.2   Nesting of capabilities within the Stock Manager agent

An agent declaration in JACK takes the form

```
agent AgentName extends Agent {
  ...
}
```

The body ('...' above) consists of declarations specifying what capabilities and plans the agent has, what beliefsets it has and what events it handles, sends and posts.

The *Stock Manager* agent has a database called *Stock Orders* and has *Stock managing* capability. Having a capability is expressed using a declaration of the form

```
#has capability CapabilityName handle;
```

The *Stock Manager* agent has a single beliefset of type *Stock Orders*, which is accessed by the capabilities and their sub-capabilities and plans. Having a beliefset is expressed using a declaration of the form

```
#private data DataType name;
```

Finally, an agent declaration contains declarations of which events the agent posts, sends and handles. The difference between posting and sending an event is that sending takes place between agents, whereas posting is internal to an agent. JACK treats messages as a type of event and thus there is no distinction between receiving a message (from another agent) and handling an (internally posted) event: both are declared using `handles event`.

We thus have the following agent declaration, which is based on Figure 10.3 (which is a copy of Figure 8.2 on page 103).

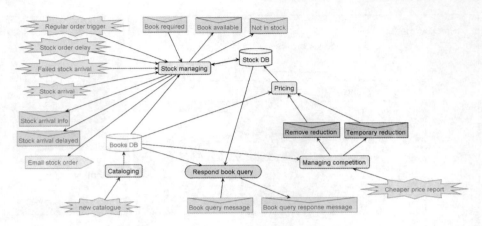

Figure 10.3   Agent overview diagram: stock manager

```
agent StockManager extends Agent {
  #has capability StockManaging sm;
  ...

  #private data Stock_Orders stockorders();
  ...

  /* Percepts */
  #handles external event Stock_order_delay;
  #handles external event Failed_stock_arrival;
  #handles external event Stock_arrival;
  #handles external event Cheaper_price_report;
  #handles external event New_Catalogue;
  ...

  /* Incoming messages */
  #handles external event Log_Outgoing_Books_Message;
  ...

  /* Outgoing message */
  #sends event Stock_arrival_info stockarrivalinfo;
  #sends event Stock_arrival_delayed stockarrivaldelayed;
  ...
}
```

10.3.2   CAPABILITIES

Capability declarations are almost identical to agent declarations. They take the form

```
capability CapabilityName extends Capability {
  ...
}
```

EXAMPLE 131

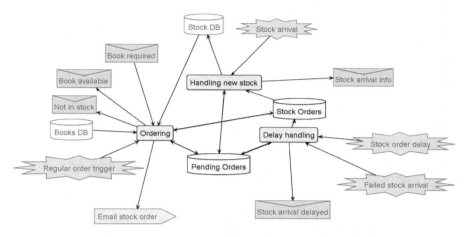

Figure 10.4   Capability diagram: stock managing capability

The body of the declaration is identical in format to that of agent declarations. In this example, the *Stock Managing* capability has *Ordering* as a sub-capability. Both capabilities access the agent's *Stock Orders* beliefset, which is declared using import data rather than private data. This indicates that the database is external to the capability. In addition to accessing the agent's *Stock Orders* database, the *Ordering* capability also has a private database of *Monthly Order*.

The following JACK code is based on Figures 10.4 and 10.5 (which are copies of Figures 9.1 and 9.2 on page 111).

```
capability Stockmanaging extends Capability {
  #has capability Ordering ordercap;
  ...

  /* import agent's databases */
  #imports data Stock_Orders stockorders();
  ...

  /* Percepts */
  #handles external event Stock_order_delay;
  #handles external event Failed_stock_arrival;
  #handles external event Stock_arrival;
  ...

  /* Incoming messages */
  #handles external event Book_required;
  ...

  /* Outgoing messages (external to agent) */
  #sends event Stock_arrival_info stockarrivalinfo;
  #sends event Stock_arrival_delayed stockarrivaldelayed;
  ...
```

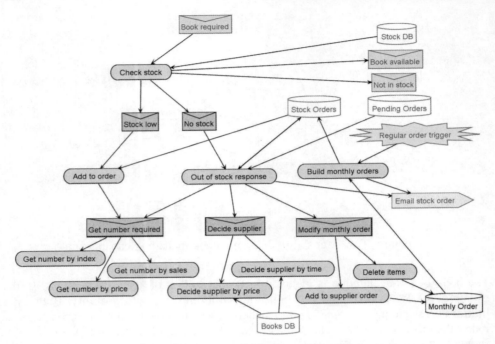

Figure 10.5   Capability overview diagram: ordering capability

```
    /* Outgoing events (internal to agent) */
    #posts event Book_available;
    #posts event Not_in_stock;
    ...
}

The following
capability Ordering extends Capability {
    /* import agent's databases */
    #imports data Stock_Orders stockorders();
    #private data Monthly_Order monthlyorder();
    ...

    /* Incoming messages */
    #handles external event Book_Required;
    ...

    /* Internal events */
    #handles event Stock_low;
    #posts event Stock_low;
    #handles event Get_number_required;
    #posts event Get_number_required;
    #handles event Decide_supplier;
```

EXAMPLE 133

```
#posts event Decide_supplier;
#handles event Modify_monthly_order;
#posts event Modify_monthly_order;
  ...

/* Plans that the capability has */
#uses plan Check_stock;
#uses plan Add_to_order;
#uses plan Get_number_by_index;
#uses plan Get_number_by_sales;
#uses plan Get_number_by_price;
  ...
}
```

10.3.3 DATA

Data can be represented in a range of ways. Data that is represented as objects can, in JACK, be simply programmed in Java.

Data that is represented as relational databases is described using the following descriptor form[24]:

Data: *name*
Description: Description of the data
Included fields/aspects: What fields does the database contain?
Key(s): Which of these fields are the key, that is, which fields are sufficient to uniquely determine a tuple.

Electronic Bookstore: Case study

Data: Stock_Orders
Description: Stores the amount of each book that is on order. The bookid is the key.
Data type: relational database
Included fields/aspects: bookid (number), amount on order (number)
Persistent: no
External to system: no

Data that is represented as relational databases can be mapped into JACK beliefsets. The *Stock Orders* database maps directly to the following JACK code.

```
beliefset Stock_Orders extends ClosedWorld {
  #key field int bookid
  #value field int amount_on_order
  #indexed query getord(int bookid, logical int amount_on_order);
}
```

---

[24]The descriptors in this chapter include information that can be found only in the descriptors. For example, in the data descriptor, the fields 'produced by' can be obtained from the diagrams and so are not included.

The `ClosedWorld` means that the database stores facts, and anything not stored is assumed to be false. An `OpenWorld` database stores both positive and negative information, and anything not stored is assumed to be unknown. Usually, `ClosedWorld` is appropriate.

The declaration of the query uses logical variables to indicate outputs. So, the declaration above automatically generates a method `getord` that takes a `bookid` and looks up the corresponding amount on order.

10.3.4 MESSAGES/EVENTS

Messages are described using the following descriptor form:

Message: *name*
Description: Description of the message
Purpose: What is the aim of the event?
Carried Information: What data does the event carry/contain?
Coverage & Overlap: Will there always be at least one applicable plan? Will there always be at most one applicable plan instance?

Electronic Bookstore: Case study

Message: Stock_low
Purpose: Cause more stock to be ordered
Carried Information: Book record, current stock held (number)
Coverage & Overlap: Covered, no overlap

Message: *Get number required*
Purpose: Determine how many copies to order
Carried Information: Book record, current stock held, number
Coverage & Overlap: not covered: There can be situations where no plan matches, if for instance not all books have a category field, overlap: *Get number by sales*, *Get number by price* and *Get number by index* can all overlap. Preference ordering is sales, price, index.

Message: *Decide supplier*
Purpose: Decide on which supplier to use
Carried Information: Book record, current stock held, amount to order
Coverage & Overlap: Covered, no overlap

Message: *Modify monthly order*
Purpose: Modify the monthly order
Carried Information: Book record, current stock held, amount to order, supplier to use
Coverage & Overlap: Covered, no overlap

Mapping these descriptors to a JACK implementation is straightforward. The JACK code for each event declares the data carried by the event and provides a posting method that creates the event.

One issue that needs to be considered in JACK is the *type* of the event: JACK provides a number of event types with different properties including `MessageEvent`

EXAMPLE 135

(for messages between agents) and BDIGoalEvent for goals. As a rough rule, an event should be of type BDIGoalEvent when there are multiple plans that can handle the event and it should be of type MessageEvent when it is being sent between agents.

The message get_number_required is not covered. This means that an attempt to determine how many copies to order could fail. This could be addressed either by requiring that any plan that posts get_number_required be prepared to fail, or by adding additional plans to handle the message. For example, one could add a plan that simply asked a clerk how many books to order. Finally, another alternative is to make (and document!) assumptions that will guarantee that the message is covered, for example, assuming that all books have categories will ensure that the message is covered.

```
event Stock_low extends BDIGoalEvent {
  BookRecord bookrecord;
  int currentstock;

  #posted as stocklow(BookRecord br, int st) {
    currentstock = st;
    bookrecord = br;
  }
}

event Get_number_required extends BDIGoalEvent {
  BookRecord bookrecord;
  int currentstock;
  int amount;

  #posted as get_number_required(BookRecord br, int st) {
    currentstock = st;
    bookrecord = br;
  }
}

event Decide_supplier extends Event {
  BookRecord bookrecord;
  int currentstock;
  int amount;
  String supplier;

  #posted as decide_supplier(BookRecord br, int st, int am) {
    currentstock = st;
    bookrecord = br;
    amount = am;
  }
}

event Modify_monthly_order extends Event {
  BookRecord bookrecord;
  int currentstock;
  int amount;
  String supplier;
```

```
  #posted as modify_monthly_order(
    BookRecord br, int stock, int am, String supp)
  {
    currentstock = stock;
    bookrecord = br;
    amount = am;
  }
}
```

10.3.5   PLANS

Finally, each plan in the Prometheus detailed design maps to a JACK plan. Plans are described using the following descriptor form:

Plan: *name*
Description: Description of the plan
Trigger: What message or percept causes the plan to execute?
Context: In what situation should this plan be applicable?
Data used & produced: What data does this plan read and/or write?
Goal: What is the goal achieved by this plan?
Failure: When might the plan fail?
Failure recovery: What needs to be done to recover from failure?
Procedure: What is the sequence of steps that the plan performs?

Electronic Bookstore: Case study

Plan: *Add to order*
Description: Order more stock of a book.
Trigger: *Stock low*
Context: Book not already ordered
Data used & produced: Reads *Stock Orders* and *Monthly Order*
Goal: Have books in stock
Failure: never fails
Failure recovery: not applicable
Procedure:
1. Determine how much stock to order (*Get number required*)
2. Determine which supplier to use (*Decide supplier*)
3. Add the order to the monthly order (*Modify monthly order*)

This plan can be implemented in JACK with the code below. The important point is that the mapping is one-to-one: each plan in the detailed design becomes a single-plan declaration in a BDI platform such as JACK. The mechanism for running plans in response to events being posted is provided by the BDI platform.

EXAMPLE 137

The information in the plan descriptor also gives much of the information in the JACK plan declaration. For example, the `handles` event declaration is the trigger in the plan descriptor and the context condition is given in pseudo-code in the descriptor.

The JACK plan declaration needs to declare the events posted and sent by the plan as well as data access. This information can be seen in the capability overview diagram where the plan appears. Finally, the plan body needs to be expressed in code rather than pseudo-code. In JACK, the language used to write plan bodies is Java extended with '@ statements' such as @achieve or @send.

The plan body below follows the four steps in the plan descriptor. Steps two, three and four are realized by posting an event as a sub-task of the current plan using @subtask. In all three cases, the processing of the plan suspends until the event either succeeds or fails. The difference between @subtask and @post is that the latter is asynchronous – it does not suspend the plan while the event is handled.

> ☞ TIP: Often, when a plan posts an event to trigger other plans, there will be some information that results from the sub-tasks. For example, posting *Decide supplier* results in a supplier being chosen. One way of communicating this information back to the parent plan is to have an additional field in the event that is set by the child plan.

```
plan Add_to_order extends Plan {
  /* Trigger */
  #handles event Stock_low stocklow;

  /* Events/messages sent */
  #posts event Get_number_required getnumber;
  #posts event Decide_supplier decidesupplier;
  #posts event Modify_monthly_order modifymonthlyorder;

  /* Data read/modified */
  #reads data Stock_Orders stockorders;
  #reads data Monthly_Order monthlyorder;

  /* Context condition */
  context() {
    stockorders.getord(stocklow.bookrecord.bookid,sorder)
        && sorder.as_int()==0 && ... ;
  }

  /* The body of the plan */
  body() {
    logical int sorder;
    /* Create event */
    getnumber = getnumber.get_number_required(
                stocklow.bookrecord, stocklow.currentstock);
    /* Post the event */
    @subtask(getnumber);
```

```
  decidesupplier = decidesupplier.decide_supplier(stocklow.bookrecord,
                     stocklow.currentstock, getnumber.amount);
  @subtask(decidesupplier);

  modifymonthlyorder = modifymonthlyorder.modify_monthly_order(
                     stocklow.bookrecord, stocklow.currentstock,
                     getnumber.amount, decidesupplier.supplier);
  @subtask(modifymonthlyorder);
  } // body()
} // plan Add_to_order
```

10.4 AUTOMATIC GENERATION OF SKELETON CODE

In this chapter, we have shown how the results of a Prometheus detailed design map naturally to (skeleton) code in an agent-oriented programming language, specifically JACK. In fact, much of this code can be automatically generated from a structured design. The JACK Intelligent Agents environment contains a tool, 'JACK Development Environment' (JDE), which supports this automatic generation of skeleton (JACK) code from a graphical design.

We have developed a prototype tool 'Prometheus Design Tool' (PDT), available at www.cs.rmit.edu.au/agents/pdt. This tool supports the process described in this book, of system specification, architectural design and detailed design. The detailed design produced by PDT can be straightforwardly converted to JDE, and consequently to JACK code. A similar approach could be used to develop plug-ins that enable PDT to produce skeleton code for a range of agent-programming systems. It is our hope that the structured methodology presented, combined with tools that support the process, will assist developers in exploring the powerful agent-programming paradigm.

A

Electronic Bookstore

This appendix provides an example of the kind of design document that can be generated from the final design artifacts, after following the Prometheus methodology. This report is generated largely automatically from the Prometheus Design Tool that supports the methodology. The design is not complete in that it does not cover all parts of the system at all levels of detail. Rather, as we have progressed the level of detail, we have narrowed the scope. Also, not all aspects are equally well developed, even within the scope we have chosen. However, there is some example of all aspects of the design, and we hope this will assist in making concrete the approach.

Also, there is no claim that this is an especially good design! There are many places where it can clearly be improved. It is developed as a way of concretely illustrating the methodology. In fact, having the methodology enables one to see and recognize where improvements can be made. If a real system was being developed, prototype implementation would also almost certainly have fed into the design. This has not happened in this case (as there was never an intent to actually build the system). Also, data descriptions are not developed in detail as it is assumed users know how to do this using alternative methodologies.

Name Electronic Bookstore
Description
Author Lin Padgham and Michael Winikoff
Version
Prometheus version 1.2
Report generation date 2003-12-08

Developing Intelligent Agent Systems   L. Padgham & M. Winikoff
© 2004 John Wiley & Sons, Ltd   ISBN: 0-470-86120-7 (HB)

1. System Specification

1.1 Goal Overview Diagram

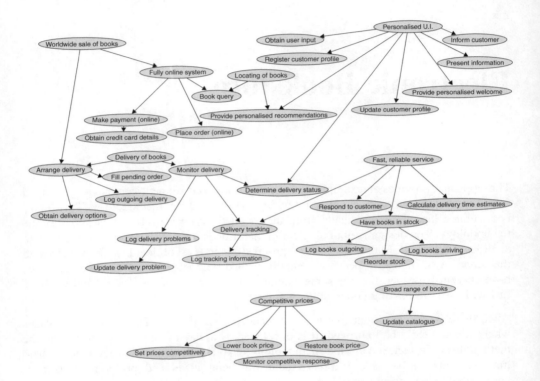

1.2 Functionalities

Functionalities Diagram

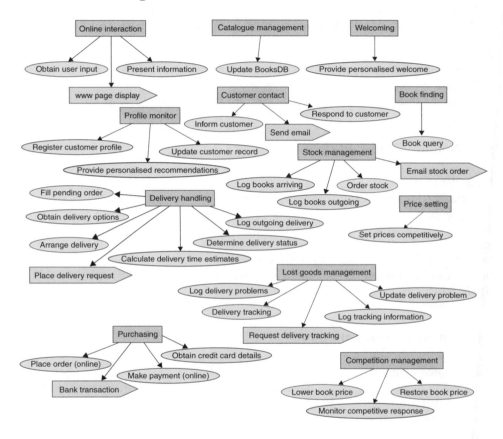

(In the functionality descriptors shown below, data which was initially specified within the functionality, but later removed due to design decisions, is shown in parentheses)

Functionality Online interaction

Name Online interaction
Description This functionality manages the online interaction with a single user, via the website.
Triggers user input
Actions WWW page display
Information used Customer DB, Customer Orders
Information produced (Customer DB, Customer Orders)
Goals Obtain user input, Present information

Functionality Catalogue management

Name Catalogue management
Description Ensures that there is an up-to-date catalogue for a wide variety of books. Updates the catalogue whenever information arrives from suppliers.
Triggers new catalogue
Actions
Information used Catalogue
Information produced Books DB
Goals Update BooksDB

Functionality Welcoming

Name Welcoming
Description Provide personalized and contextualized welcome messages when a user logs into the WWWsite.
Triggers Arrival at WWWsite
Actions
Information used Customer DB, Customer Orders
Information produced Temporary WWWpage data, (Customer DB)
Goals Provide personalized welcome

Functionality Profile monitor

Name Profile monitor
Description manages the user's personal profile and any information contributing to or based on that profile
Triggers Update customer profile, Register customer profile, Request profile information
Actions
Information used Customer DB
Information produced Customer DB
Goals Provide personalized recommendations, Register customer profile, Update customer record

Functionality Stock management

Name Stock management
Description Keeps track of stock on hand, ordering stock as required and monitoring delivery of those orders.
Triggers Failed stock arrival, Stock arrival, Stock order delay
Actions Email stock order
Information used Stock DB, Books DB, Stock Orders, Pending Orders

Information produced Stock DB, Stock Orders, Pending Orders
Goals Log books arriving, Log books outgoing, Order stock

Functionality Book finding

Name Book finding
Description Locates book or book information according to specification. (Currently limited to searching books Searches for books in the Stock DB and in the Books DB, but could be extended to search pro-actively for books not in these sources.)
Triggers Book query
Actions
Information used Stock DB, Books DB
Information produced Temporary booklist
Goals Book query

Functionality Delivery handling

Name Delivery handling
Description This functionality manages delivery of orders to customers
Triggers
Actions Place delivery request
Information used Courier DB, Postal DB, Customer Orders
Information produced Customer Orders
Goals Arrange delivery, Calculate delivery time estimates, Determine delivery status, Log outgoing delivery, Obtain delivery options, Fill pending order

Functionality Lost goods management

Name Lost goods management
Description Manages queries about books that have not arrived. Does tracking and arranges duplicate books if needed.
Triggers Tracking info, No tracking response
Actions Request delivery tracking
Information used Delivery Problems
Information produced Delivery Problems, (Customer DB, Customer Orders)
Goals Log delivery problems, Delivery tracking, Log tracking information, Update delivery problem

Functionality Price setting

Name Price setting
Description Set prices competitively, based on cost of books. (Further development could allow for mark-up to be based also on the popularity of the title. It could also allow for pro-active checking of prices elsewhere.)
Triggers New catalogue, Temporary reduction, Remove reduction

Actions
Information used Stock DB, Books DB
Information produced Stock DB
Goals Set prices competitively

Functionality Purchasing

Name Purchasing
Description Manages on-line sales of books, including credit card transaction.
Triggers Bank transaction response
Actions Bank transaction
Information used Customer Orders, Customer DB
Information produced Purchase approval (temporary) (Customer DB, Customer Orders)
Goals Place order (online), Make payment (online), Obtain credit card details

Functionality Competition management

Name Competition management
Description Lowers prices temporarily if needed to stay competitive. Monitors situation and restores prices when possible.
Triggers Cheaper price report
Actions
Information used Books DB, Stock DB
Information produced Stock DB
Goals Lower book price, Monitor competitive response, Restore book price

Functionality Customer contact

Name Customer contact
Description contacts customer with information on delivery details as needed. Contact point for customer queries regarding orders.
Triggers Book delivery information, response to Late delivery query
Actions Send email
Information used Customer DB, Customer Orders
Information produced
Goals Inform customer, Respond to customer

1.3 Scenarios

Scenario Query late books

Name Query late books
Description This scenario illustrates what happens when a user enquires about books ordered which have not arrived. Information is obtained from delivery records and as the

book is delayed, a tracking process is started. The customer is notified. After a wait, the book cannot be located and a new book is sent.

Trigger Late delivery query

Steps

#	Type	Name	Functionality	Data
1	Goal	Determine delivery status	Delivery handling	uses: Customer Orders
2	Goal	Log delivery problems	Lost goods management	produces: Delivery Problems
3	Action	Request delivery tracking	Lost goods management	uses: Customer Orders, Courier DB produces: tracking request
4	Goal	Inform customer	Customer contact	uses: Customer DB
5	Other		Wait	
6	Percept	Tracking info	Lost goods management	
7	Goal	Arrange delivery	Delivery handling	uses: Customer Orders produces: Delivery Problems
8	Goal	Log books outgoing	Stock management	uses: Customer Orders produces: Stock DB
9	Goal	Inform customer	Customer contact	uses: Customer DB
10	Goal	Update delivery problem	Lost goods management	produces: Delivery Problems

Scenario Book finding scenario

Name Book finding scenario

Description Finds book(s) as requested by the user and displays the result.

Trigger user request

Steps

#	Type	Name	Functionality	Data
1	Goal	Book query	Book finding	uses: Stock DB, Books DB
2	Goal	Present information	Online interaction	uses: temporary booklist
3	Action	WWW page display	Online interaction	

Variation No books found that match request. Provide message and suggest changes to request.

Scenario Order book

Name Order book
Description An order is received from the WWW page interface (goal Place order)
Information is obtained in order to place the order and the order is placed.
Trigger Goal: Place order
Steps

#	Type	Name	Functionality	Data
1	Goal	Obtain delivery options	Delivery handling	uses: Courier DB, Postal DB
2	Goal	Calculate delivery time estimates	Delivery handling	customer order record, Courier DB, Postal DB produces: time estimates
3	Goal	Present information	Online interaction	uses: temporary data
4	Percept	user input	Online interaction	Get delivery choice
5	Goal	Obtain credit card details	Purchasing	uses: Customer DB
6	Percept	user input	Online interaction	uses: Customer DB produces: CC details (if needed), agreed transaction
7	Action	Bank transaction	Purchasing	transaction details (temp)
8	Percept	Bank transaction response	Purchasing	
9	Goal	Arrange delivery	Delivery handling	produces: Customer Orders, uses and produces: customer order record
10	Action	Place delivery request	Delivery handling	uses: customer order record
11	Goal	Log outgoing delivery	Delivery handling	produces: Customer Orders
12	Goal	Log books outgoing	Stock management	uses: customer order record produces: Stock DB
13	Goal	Update customer record	Profile monitor	produces: Customer DB
14	Action	Send email	Customer contact	uses: Customer DB

Variation 1: Book is not currently available. Include information with delivery options.
Replace steps 7-12 with steps to add the order to an orders pending file.

Scenario Pending order arrives

Name Pending order arrives
Description Stock arrives that is needed for a pending order. Delivery is arranged, internal data updated and customer notified.
Trigger Message that book for a pending order has arrived.
Steps

#	Type	Name	Functionality	Data
1	Percept	Stock arrival	Stock management	
2	Goal	Fill pending order	Delivery handling	produces: Pending Orders
3	Goal	Arrange delivery	Delivery handling	produces: Customer Orders, uses and produces: customer order record
4	Action	Place delivery request	Delivery handling	uses: customer order record
5	Goal	Inform customer	Customer contact	uses: Customer DB, customer order record
6	Goal	Log books outgoing	Stock management	uses: customer order record produces: Stock DB
7	Goal	Log outgoing delivery	Delivery handling	produces: Customer Orders

Variation Waiting on additional books. Book is held aside.

Scenario Stock arrival

Name Stock arrival
Description Fills pending orders and updates stock information
Trigger Stock arrival percept.
Steps

#	Type	Name	Functionality	Data
1	Goal	Fill pending order	Delivery handling	uses and produces: Pending Orders
2	Goal	Log outgoing delivery	Delivery handling	produces: Customer Orders
3	Goal	Log books outgoing	Stock management	uses: customer order record produces: Stock DB

Variation No pending orders to fill. Leave out those steps.

Scenario Stock delayed

Name Stock delayed
Description Supplier notifies delay. Check pending orders delayed. Notify customer contact, who may notify customer.
Trigger message from supplier.
Steps

#	Type	Name	Functionality	Data
1	Goal	Identify affected orders	Stock management	uses and produces: Pending Orders
2	Goal	Inform customer	Customer contact	uses and produces: Customer DB

Scenario Stock order

Name Stock order
Description Order made and no stock available, so order is placed. When stock arrives, order is filled and sent.
Trigger order made and no stock available.
Steps

#	Type	Name	Functionality	Data
1	Goal	Order stock	Stock management	produces: Stock Orders
2	Action	Email stock order	Stock management	order list (temp)
3	"Other"		wait for stock to	
4	Scenario	Stock arrival		

Variation Stock may not arrive, in which case replace Stock arrival scenario by Missed stock arrival scenario.

A message may come saying that stock arrival is delayed, in which case replace Stock arrival scenario by Stock delayed scenario.

Various other stock ordering scenarios. Stock drops below a threshold and is placed on an order list. Regular data-based order schedule.

Scenario Order status query

Name Order status query
Description A request is received to review order status. Information is obtained and presented to the user.
Trigger user request received via WWW page

Steps

#	Type	Name	Functionality	Data
1	Goal	Determine delivery status	Delivery handling	uses: Customer Orders, Pending Orders
2	Goal	Present information	Online interaction	uses: temporary message
3	Goal	Update customer record	Profile monitor	produces: Customer DB

Scenario Customer profile update

Name Customer profile update
Description update the user's profile at user request
Trigger trigger: User input specifying update.
Steps

#	Type	Name	Functionality	Data
1	Goal	Confirm changes	Online interaction	uses: Customer DB produces: customer profile record
2	Goal	Update customer record	Customer contact	uses: updated info (temp) produces: Customer DB

Variation User doesn't confirm changes. Leave unchanged–i.e. don't do step 2.

Scenario WWWsite arrival

Name WWWsite arrival
Description New customer arrives. Welcome page displayed.
Trigger new arrival at site.
Steps

#	Type	Name	Functionality	Data
1	Goal	Provide personalized welcome	Welcoming	uses: Customer DB
2	Action	WWW page display	Online interaction	uses: welcome text (temp)

Variation Customer nor previously registered. Add initial steps to obtain profile and register customer.

Scenario Cheaper price notification

Name Cheaper price notification
Description Cheaper price from competitor notified. System price is temporarily adjusted, monitored, and then reset to normal when possible.

Trigger cheaper price report (percept)
Steps

#	Type	Name	Functionality	Data
1	Goal	Lower book price	Competition management	uses: Books DB produces: Stock DB
2	Goal	Monitor competitive response	Competition management	
3	Goal	Restore book price	Competition management	produces: Stock DB

Scenario New catalogue

Name New catalogue
Description New book catalogue arrives from supplier. Books DB is updated.
Trigger New catalogue (percept)
Steps

#	Type	Name	Functionality	Data
1	Goal	Update BooksDB	Catalogue management	produces: Books DB

Scenario Missed stock arrival

Name Missed stock arrival
Description Stock does not arrive when due. No information available from supplier. Contact supplier for info then notify delay.
Trigger Trigger: Arrival time passes and no stock arrival.
Steps

#	Type	Name	Functionality	Data
1	Action	Send email	Stock management	uses: Stock Orders
2	"Other"	Stock management	wait for response re delay	
3	Scenario	Stock delayed		

Variation No response from supplier in given time. Do stock delayed scenario and keep trying to contact.

1.4 Percepts

Percept user input

Name user input
Description user input from WWWsite.
Defined: Page 156

Percept Arrival at WWWsite

Name Arrival at WWWsite
Description Indication of a new arrival at the WWWsite
Defined: Page 157

Percept Bank transaction response

Name Bank transaction response
Description Response to request for credit card payment
Defined: Page 157

Percept Tracking info

Name Tracking info
Description response from courier company to a tracking request
Defined: Page 157

Percept new catalogue

Name new catalogue
Description New catalogue information from a supplier
Defined: Page 157

Percept Stock arrival

Name Stock arrival
Description information that stock has arrived
Defined: Page 158

Percept Failed stock arrival

Name Failed stock arrival
Description Indication that a stock order has not arrived when expected.
Defined: Page 158

Percept Stock order delay

Name Stock order delay
Description Information from supplier regarding delivery delay
Defined: Page 158

Percept Cheaper price report

Name Cheaper price report

Description External report indicating a supplier selling a book at a cheaper price than this company.
Defined: Page 158

Percept Regular order trigger

Name Regular order trigger
Description Timer-based trigger for placing regular stock orders
Defined: Page 159

Percept No tracking response

Name No tracking response
Description Percept generated by system monitor if no response received after a set time.
Defined: Page 159

1.5 Actions

Action Email stock order

Name Email stock order
Description Action to order stock from a supplier
Defined: Page 159

Action Request delivery tracking

Name Request delivery tracking
Description send request to courier or postal service to track an item which has not arrived.
Defined: Page 160

Action Place delivery request

Name Place delivery request
Description Send an email request for delivery pick-up, either by courier, or by the postal room.
Defined: Page 160

Action Send email

Name Send email
Description Send email message (generic)
Defined: Page 160

Action WWW page display

Name WWW page display
Description Displays WWW page content
Defined: Page 160

Action Bank transaction

Name Bank transaction
Description Action which executes a bank transaction
Defined: Page 160

1.6 Data

External

Courier DB
Contains information about courier companies, areas and rates.

Postal DB
Contains information about postal rates.

Other

Customer DB
contains information about customers, their profile, their history of visits to the site and orders, etc.

Customer Orders
contains records with orders that have been (fairly recently) sent.

Pending Orders
contains records with orders which have been placed, but not yet sent.

Delivery Problems
contains records of queries about nonarriving items and the investigation of such.

Books DB
contains a comprehensive listing of books, with information on suppliers, prices, etc. Not all books are necessarily stocked.

Stock DB
contains records of books that are stocked.

Stock Orders
contains records of stock orders placed and awaiting delivery.

2. Architectural Design

2.1 System Overview Diagram

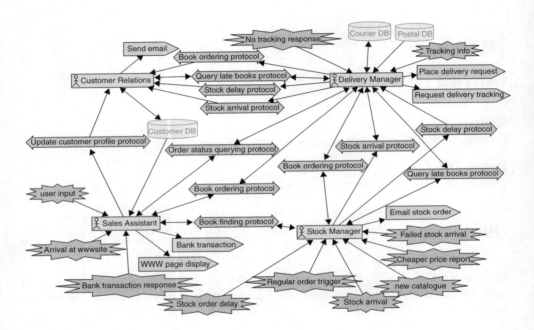

2.2 Agents

Agent Sales Assistant

Name Sales Assistant
Description This agent deals with all online interaction with a customer - analogous to a shop assistant in a physical store. This includes helping the customer find appropriate books as well as passing on other enquiries such as information update or order tracking requests.

Cardinality minimum 0
Cardinality maximum many
Lifetime while a particular user is at the WWWsite
Initialization Data from Customer DB
Demise Ensures that customer profile information is passed on to Customer Relations. Releases all temporary data structures. Closes any communications lines.
Percepts user input, Arrival at WWWsite, Bank transaction response
Actions WWW page display, Bank transaction
Uses data Customer DB
Produces data customer order record
Internal data not yet defined
Goals Locating of books, Make payment (online), Obtain credit card details, Obtain delivery options, Obtain user input, Personalized U.I., Present information, Provide personalized welcome, Register customer profile, Update customer record
Functionalities Book finding, Online interaction, Purchasing, Welcoming
Protocols Order status querying protocol, Update customer profile protocol, Book finding protocol, Book ordering protocol

Agent Delivery Manager

Name Delivery Manager
Description Arranges all aspects of delivery to customer. Deals with any problems with deliveries, including notifying customer relations agent of issues that affect customers.
Cardinality minimum 1
Cardinality maximum 1
Lifetime ongoing
Initialization
Demise Write out all internal data structures.
Percepts No tracking response, Tracking info
Actions Request delivery tracking, Place delivery request
Uses data Postal DB, Courier DB
Produces data Postal DB, Courier DB
Internal data Customer Orders
Goals Fill pending order, Obtain delivery options, Arrange delivery, Calculate delivery time estimates, Determine delivery status, Log outgoing delivery, Log delivery problems, Delivery tracking, Log tracking information, Update delivery problem
Functionalities Delivery handling, Lost goods management
Protocols Query late books protocol, Order status querying protocol, Stock arrival protocol, Stock delay protocol, Book ordering protocol

Agent Customer Relations

Name Customer Relations
Description This agent deals with all offline interaction with the customer (e.g. sending updates on orders, etc.) as well as maintenance of the DB of customer information and profiles.

Cardinality minimum 1
Cardinality maximum 1
Lifetime ongoing
Initialization
Demise None
Percepts
Actions Send email
Uses data Customer DB
Produces data Customer DB
Internal data
Goals Inform customer, Provide personalized recommendations, Register customer profile, Respond to customer, Update customer record
Functionalities Customer contact, Profile monitor
Protocols Query late books protocol, Update customer profile protocol, Stock arrival protocol, Stock delay protocol, Book ordering protocol

Agent Stock Manager

Name Stock Manager
Description Deals with all aspects of books available from store. Includes ensuring that books are available, pricing, reordering, monitoring deliveries, etc.
Cardinality minimum 1
Cardinality maximum 1
Lifetime
Initialization
Demise N/A. All data written out on an ongoing basis.
Percepts Stock order delay, new catalogue, Cheaper price report, Failed stock arrival, Stock arrival, Regular order trigger
Actions Email stock order
Uses data
Produces data
Internal data Books DB, Stock Orders, Stock DB
Goals Log books outgoing, Log books arriving, Order stock, Lower book price, Monitor competitive response, Restore book price, Set prices competitively, Monitor stock arrivals, Manage new stock
Functionalities Catalogue management, Competition management, Stock management, Price setting
Protocols Query late books protocol, Book finding protocol, Stock arrival protocol, Stock delay protocol, Book ordering protocol

2.3 Percepts

Percept user input

Name user input

Description user input from WWWsite.
Information carried Selection of item, accompanying field values
Knowledge updated
Source WWWpage
Processing WWWpage software translates mouseclicks and location to symbolic items.
Agents responding Sales Assistant
Expected frequency Can be 1–2 per second.

Percept Arrival at WWWsite

Name Arrival at WWWsite
Description Indication of a new arrival at the WWWsite
Information carried Customer ID (if cookie available)
Knowledge updated Customer visit history
Source WWWsite listener
Processing Extraction of relevant information from cookie.
Agents responding Sales Assistant
Expected frequency No higher than 10 per minute

Percept Bank transaction response

Name Bank transaction response
Description Response to request for credit card payment
Information carried Accept/Reject, fraud (optional), amount, account ID
Knowledge updated none
Source bank processing system
Processing none
Agents responding Sales Assistant
Expected frequency Individual agent unlikely to receive more than 1 in total. Certainly no more than 1 every few minutes, maximum. System as a whole could potentially receive around 10 per minute maximum.

Percept Tracking info

Name Tracking info
Description response from courier company to a tracking request
Information carried Tracking request ID, Located/not located, damaged/undamaged
Knowledge updated courier reliability, problem record
Source courier company input
Processing Structured input system, processing for correct values etc.
Agents responding Delivery Manager
Expected frequency infrequent

Percept new catalogue

Name new catalogue
Description New catalogue information from a supplier

Information carried Book information, pricing, release dates, etc.
Knowledge updated Books DB records
Source supplier interface system
Processing Parsing of information and extraction of relevant fields to standard format.
Agents responding Stock Manager
Expected frequency No more than 10/month

Percept Stock arrival

Name Stock arrival
Description information that stock has arrived
Information carried Stock order ID, item list, supplier ID
Knowledge updated Supplier reliability
Source External system
Processing Fields extracted from input received from external system.
Agents responding
Expected frequency Approximately 10 per month. Never closer than a few minutes.

Percept Failed stock arrival

Name Failed stock arrival
Description Indication that a stock order has not arrived when expected.
Information carried Stock order ID
Knowledge updated Reliability of supplier, expected date of arrival of order.
Source System monitor, using system time.
Processing Checking that order has not in fact arrived and also whether any information has been received.
Agents responding Stock Manager
Expected frequency Infrequent

Percept Stock order delay

Name Stock order delay
Description Information from supplier regarding delivery delay
Information carried stock order ID, supplier ID, previous expected delivery date, new expected delivery date, reason.
Knowledge updated Supplier reliability, expected delivery date
Source supplier input system.
Processing none
Agents responding Stock Manager
Expected frequency infrequent.

Percept Cheaper price report

Name Cheaper price report
Description External report indicating a supplier selling a book at a cheaper price than this company.

Information carried Book ID, price, competitor information, date, information source.
Knowledge updated Competition records
Source Human interface
Processing None
Agents responding Stock Manager
Expected frequency Relatively seldom. Could be several per minute at a particular time (when person is inputting information).

Percept Regular order trigger

Name Regular order trigger
Description Timer-based trigger for placing regular stock orders
Information carried none
Knowledge updated none
Source system monitor attached to system clock.
Processing none
Agents responding Stock Manager
Expected frequency Once every set period, possibly monthly.

Percept No tracking response

Name No tracking response
Description Percept generated by system monitor if no response received after a set time.
Information carried tracking request ID
Knowledge updated courier reliability records, delivery problem record
Source monitor attached to system clock.
Processing none

Agent Sales Assistant

Agents responding Delivery Manager
Expected frequency Infrequent

2.4 Actions

Action Email stock order

Name Email stock order
Description Action to order stock from a supplier
Parameters supplier email address, order list, urgency
Duration Immediate (results take time but order action is immediate)
Failure May receive bounced email message at a later time. Failure can go unnoticed.
Partial change None
Side effects None

Action Request delivery tracking

Name Request delivery tracking
Description send request to courier or postal service to track an item which has not arrived.
Parameters Courier or postal service email address, package ID, date package sent, address on package
Duration Immediate
Failure May bounce or not arrive.
Partial change None
Side effects None

Action Place delivery request

Name Place delivery request
Description Send an email request for delivery pick-up, either by courier, or by the postal room.
Parameters email address (postal room, courier company, etc), delivery address, goods list
Duration Immediate (action to send request is immediate - results will take time)
Failure Mail may bounce. May also not be received without visible bounce.
Partial change None
Side effects None

Action Send email

Name Send email
Description Send email message (generic)
Parameters address, content, sender address
Duration Immediate
Failure May bounce or fail to arrive.
Partial change N/A
Side effects None

Action WWW page display

Name WWW page display
Description Displays WWW page content
Parameters Content, terminal type
Duration Durational - may take several seconds.
Failure Crash of browser process, hanging of browser process.
Partial change N/A
Side effects None

Action Bank transaction

Name Bank transaction
Description Action which executes a bank transaction

Parameters Amount, bank account number, transaction-type
Duration Durational, normally a few seconds
Failure Usually failure message received if failure experienced
Partial change None
Side effects None

2.5 Protocols

Protocol Book finding protocol

Name Book finding protocol
Description Interaction around a user query regarding (a) book(s)
Included messages Book query message, Book query response message
Scenarios Book finding scenario
Agents Sales Assistant, Stock Manager
Notes

Protocol Update customer profile protocol

Name Update customer profile protocol
Description Protocol for updating of customer profile.
Included messages Update customer profile message, Register new customer message
Scenarios Customer profile update
Agents Sales Assistant, Customer Relations
Notes

Protocol Book ordering protocol

Name Book ordering protocol
Description Interactions as a result of customer placing an order.
Included messages Get delivery information message, Delivery options information, Book purchase, Book required, Not in stock, Book available, Book delivery information
Scenarios Order book
Agents Sales Assistant, Delivery Manager, Stock Manager, Customer Relations
Notes

Protocol Order status querying protocol

Name Order status querying protocol
Description Interaction following a customer enquiry about order status
Included messages Determine delivery status message, Determine delivery status reply message
Scenarios order status query
Agents Sales Assistant, Delivery Manager
Notes

Protocol Query late books protocol

Name Query late books protocol
Description Interaction following customer enquiry about late arriving books.
Included messages Late delivery query, Book expected soon, Investigating, Book required, Not in stock, Book available, Book delivery information, Book located,
Scenarios Query late books scenario
Agents Customer Relations, Sales Assistant, Delivery Manager
Notes

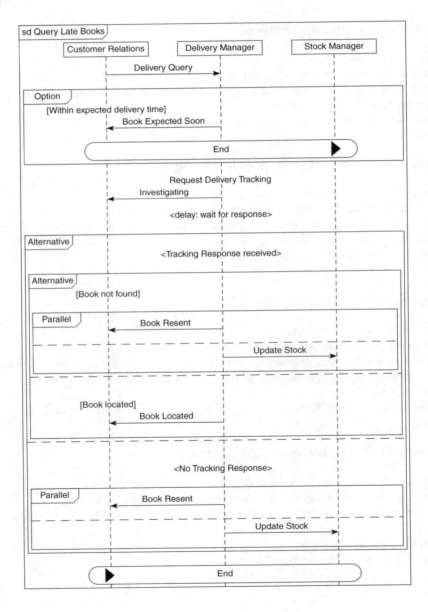

Protocol Stock arrival protocol

Name Stock arrival protocol
Description Interaction resulting from arrival of new stock. Includes filling of pending orders.
Included messages Stock arrival info, Book delivery information,
Scenarios stock arrival, Pending order arrives
Agents Stock Manager, Delivery manager, Customer Relations
Notes Note that I have planned to include pending order arrived scenario within this (maybe)–else add pending order arrived.

Protocol Stock delay protocol

Name Stock delay protocol
Description Interaction resulting from delayed stock arrival.
Included messages Stock arrival delayed, Book delivery information
Scenarios Missed stock arrival, Stock delayed
Agents Stock Manager, Delivery Manager, Customer Relations
Notes note this includes both notified delay, and also just not turning up. Two different triggers.

2.6 Messages

Message Determine delivery status message

Name Determine delivery status message
Description Message to obtain information on delivery status of an order
Distribution Sales Assistant → Delivery Manager
Purpose To obtain information for presentation to an online user.
Carried information Order ID

Message Late delivery query

Name Late delivery query
Description Message querying books that are late and not arrived.
Distribution Sales Assistant → Delivery Manager, Customer Relations → Delivery Manager
Purpose To obtain information, or to start a tracking process.
Carried information Order ID

Message Book delivery information

Name Book delivery information
Description Information sent to customer relations regarding the delivery details of a book that has been ordered.

Distribution Delivery Manager → Customer Relations
Purpose To allow customer Relations to notify customer that the book has been sent.
Carried information date sent, order ID, customer ID

Message Determine delivery status reply message

Name Determine delivery status reply message
Description Status information regarding an order
Distribution Delivery Manager → Sales Assistant
Purpose To provide information to be given to customer online.
Carried information List of items, date sent, address sent to, method of delivery, expected arrival date.

Message Update customer profile message

Name Update customer profile message
Description Message to update a customer's profile
Distribution Sales Assistant → Customer Relations
Purpose To pass information for updating to Customer Relations agent
Carried information customer profile record

Message Book query message

Name Book query message
Description Query regarding a book, formulated from user request.
Distribution Sales Assistant → Stock Manager
Purpose To be used in querying the Stock DB and/or the Books DB to return a response to the user.
Carried information A well formulated query.

Message Book query response message

Name Book query response message
Description Response from book query: a list of book records.
Distribution Stock Manager → Sales Assistant,
Purpose For use in displaying a response to the user.
Carried information Book records, availability.

Message Stock arrival info

Name Stock arrival info
Description Indicates that stock has arrived and lists the new stock.
Distribution Stock Manager → Delivery Manager
Purpose To allow Delivery Manager to process pending orders
Carried information Book IDs and quantity of newly arrived stock

Message Stock arrival delayed

Name Stock arrival delayed
Description Indicates that an order of stock is delayed
Distribution Stock Manager → Delivery Manager
Purpose To inform Delivery Manager so that pending orders can be checked and customers notified if necessary.
Carried information Stock order (Book IDs), expected delivery date

Message Book expected soon

Name Book expected soon
Description Response from delivery manager, following a query about books that have not arrived.
Distribution Delivery Manager → Customer Relations
Purpose To allow response to customer regarding query.
Carried information Date order sent. Also customer ID, order ID.

Message Investigating

Name Investigating
Description Message indicating that late delivery is being investigated.
Distribution Delivery Manager → Customer Relations
Purpose To provide information.
Carried information Order details, date sent, date should have arrived.

Message Book located

Name Book located
Description Message resulting from tracking of books that have not arrived.
Distribution Delivery Manager → Customer Relations
Purpose To enable Customer Relations agent to tell customer that books have been located and should arrive shortly.
Carried information Date, Order ID, Customer ID

Message Get delivery information message

Name Get delivery information message
Description Request for information on delivery options to an area, for a particular order.
Distribution Sales Assistant → Delivery Manager
Purpose To obtain information from Delivery manager on delivery options.
Carried information Order list, address.

Message Delivery options information

Name Delivery options information

Description Message outlining delivery options and pricing for a particular area and set of books.
Distribution Delivery Manager $\rightarrow$ Sales Assistant
Purpose To provide information which will be given to the user for selection of preferred option.
Carried information Options, each containing price and time estimate.

Message Book purchase

Name Book purchase
Description Information required by delivery manager for arranging delivery of the book(s)
Distribution Sales Assistant $\rightarrow$ Delivery Manager
Purpose To provide information for delivery and to trigger arranging of the delivery.
Carried information Book(s) ordered, delivery address, customer ID, delivery method chosen

Message Register new customer message

Name Register new customer message
Description Message to register a new customer.
Distribution Sales Assistant $\rightarrow$ Customer Relations
Purpose To provide information and trigger adding the customer to the Customer DB.
Carried information Customer profile record.

Message Book required

Name Book required
Description Book ID required for delivery to customer.
Distribution Delivery Manager $\rightarrow$ Stock Manager
Purpose To establish if book is available, and if so to notify Stock Manager to reduce quantity on hand.
Carried information Book ID, quantity.

Message Not in stock

Name Not in stock
Description Text string response message
Distribution Stock Manager $\rightarrow$ Delivery Manager
Purpose Indicator that book requested cannot be supplied. Possible response to Book required message.
Carried information none

Message Book available

Name Book available
Description Response to 'Book Required' message. Indicates book is in stock.

Distribution Stock Manager $\rightarrow$ Delivery Manager
Purpose Indicates that arranging of delivery can proceed.
Carried information none

3. Detailed Design

3.1 Agent Stock Manager

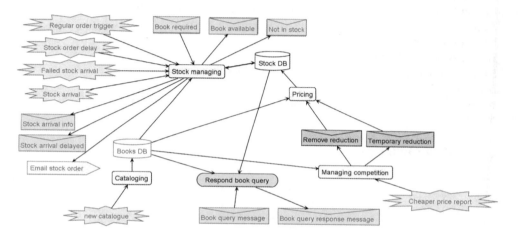

3.1.1 Capabilities

Capability Stock managing

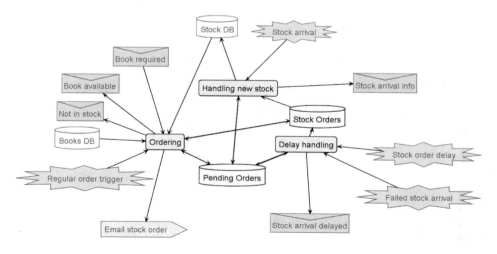

Name Stock managing
Description Ensures that stock levels are satisfactory, either by placing regular orders, or by placing immediate orders if stock runs out and the book is ordered.
Goals Order stock, Log books outgoing, Log books arriving
Processes
Protocols Stock arrival, Stock delay, Book ordering
Incoming messages Book required
Outgoing messages Stock arrival info, Stock arrival delayed, Book available, Not in stock
Internal messages
Percepts Stock order delay, Failed stock arrival, Stock arrival, Regular order trigger
Actions Email stock order
Data used: Imported Books DB
Data produced: Exported
Data internal Stock DB, Stock Orders, Pending Orders
Included plans
Included capabilities Delay handling, Ordering, Handling new stock
Notes

Capability Ordering

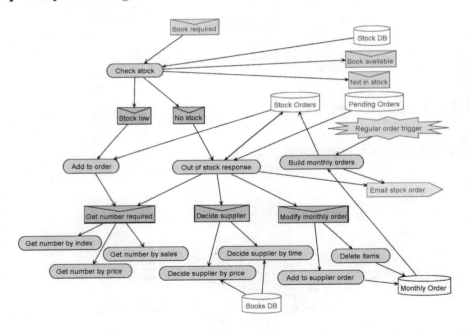

Name Ordering
Description This capability manages the ordering of stock from suppliers–either on a regular basis, or, if necessary, when stock runs out.
Goals Order stock

Processes Stock maintenance, (further not yet defined)
Protocols Book ordering protocol
Incoming messages Book required
Outgoing messages Book available, Not in stock,
Internal messages Modify monthly order, Decide supplier, Get number required, No stock, Stock low
Percepts Regular order trigger
Actions Email stock order
Data used: Imported Pending Orders, Books DB
Data produced: Exported Stock Orders, Pending Orders
Data Internal Monthly Order, (not further developed)
Included plans Check stock, Add to order, Out of stock response, Get number by index, Decide supplier by price, Decide supplier by time, Get number by price,
Get number by sales, Add to supplier order, Delete items, Build monthly orders
Included capabilities None
Notes

Capability Handling new stock

This capability is not yet developed. Descriptor results from external information.

Name Handling new stock
Description Manages arrival of new stock, updating Stock DB and providing information for any pending orders.
Goals Manage new stock
Processes Stock arrival, (not further defined)
Protocols Book ordering
Incoming messages
Outgoing messages Stock arrival info
Internal messages not yet defined
Percepts Stock arrival
Actions
Data used: Imported Pending Orders, Stock Orders
Data produced: Exported Stock DB, Pending Orders
Data internal not yet defined
Included plans not yet developed
Included capabilities none
Notes

Capability Delay handling

This capability is not yet developed. Descriptor results from external information.

Name Delay handling
Description Manage the effects of delays in arrival of books.
Goals Monitor stock arrivals
Processes not yet defined
Protocols Stock delay
Incoming messages
Outgoing messages Stock arrival delayed

Internal messages not yet defined
Percepts Stock order delay, Failed stock arrival
Actions
Data used: Imported Pending Orders
Data produced: Exported Stock Orders, Pending Orders
Data internal Not yet defined
Included plans
Included capabilities
Notes

Capability Cataloging

This capability is not yet developed. Descriptor results from external information.

Name Cataloging
Description Updates the Books DB with information from catalogues or other information sources.
Goals Update BooksDB
Processes Not yet defined
Protocols None
Incoming messages None
Outgoing messages None
Internal messages Not yet defined
Percepts new catalogue
Actions
Data used: Imported
Data produced: Exported Books DB
Data internal Not yet defined
Included plans Not yet defined
Included capabilities
Notes

Capability Managing competition

This capability is not yet developed. Descriptor results from external information.

Name Managing competition
Description Sets book prices competitively. Maintains information about competitor prices and makes temporary reductions as necessary to maintain lowest price.
Goals Competitive prices, Lower book price, Restore book price, Set prices competitively
Processes Not yet defined
Protocols None
Incoming messages
Outgoing messages Temporary reduction, Remove reduction,
Internal messages Not yet defined
Percepts Cheaper price report
Actions

Data used: Imported Books DB
Data produced: Exported
Data internal Not yet defined
Included plans Not yet defined
Included capabilities
Notes

Capability Pricing

This capability is not yet developed. Descriptor results from external information.

Name Pricing
Description Manages pricing of books in Stock DB, usually from the catalogue, but possibly also as a result of a notification due to low competitor prices.
Goals Set prices competitively
Processes Not yet defined
Protocols None
Incoming messages Temporary reduction Remove reduction,
Outgoing messages
Internal messages Not yet defined
Percepts
Actions
Data used: Imported Books DB
Data produced: Exported Stock DB
Data internal Not yet defined
Included plans Not yet defined
Included capabilities
Notes

3.1.2 Processes

Process Stock Arrival process

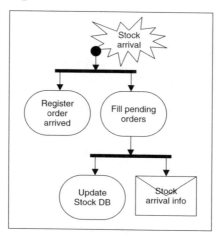

Name Stock Arrival process
Description describes the process of updating the Stock DB and notifying arrival of books for pending orders, when new stock arrives.
Activities Fill pending orders, Register order arrived, Update Stock DB
Triggers Stock arrival
Messages <Stock arrival info, to Delivery Manager>
Protocols Stock arrival protocol
Capabilities Stock Management.
Notes:

Stock Maintenance process

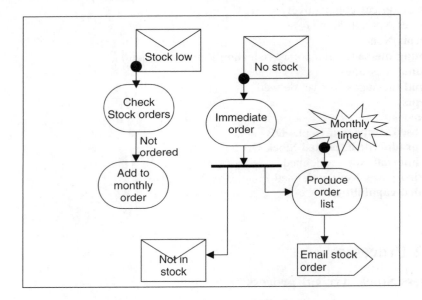

Name Stock maintenance
description The activity whereby there is an attempt to maintain sufficient stock to immediately fill orders. The activity is responsive to immediate demands as well as maintaining stock levels from month to month.
Triggers Book required, Monthly timer
Activities Check Stock DB, Immediate order, Add to monthly order, Produce order list.
Messages <Update customer profile, to Customer Relations Agent>, <Book purchase, to Delivery Manager Agent>
Protocols Order book protocol, Stock Arrival Protocol, Query Late Order
Capabilities Stock Management.
Notes

Further processes not yet defined

3.1.3 External Messages

Message Book required

Defined Page 166
Description Book ID required for delivery to customer.
Additional fields:
Coverage and overlap Full coverage, no overlap
Information carried: fields and value types/ranges:
<Book ID; ID number>

Message Book available

Defined Page 166
Description Response to 'Book Required' message. Indicates book is in stock.
Additional fields:
Coverage and overlap Full coverage, no overlap
Information carried: fields and value types/ranges:
<Book ID; ID number>

Message Not in stock

Defined Page 166
Description Text string response message
Additional fields:
Coverage and overlap Full coverage, no overlap
Information carried: fields and value types/ranges:
<Book ID; ID number>, <Expected arrival; date>

Message Stock arrival info

Defined Page 164
Description Indicates that stock has arrived and lists the new stock.
Additional fields:
Coverage and overlap Full coverage, no overlap
Information carried: fields and value types/ranges:
<Booklist; list of ID numbers>

Message Stock arrival delayed

Defined Page 165
Description Indicates that an order of stock is delayed
Additional fields:
Coverage and overlap Full coverage, no overlap
Information carried: fields and value types/ranges:
<Booklist; list of ID numbers>

Message Book query message

Defined Page 164
Description Query regarding a book, formulated from user request.
Additional fields:
Coverage and overlap Full coverage, no overlap
Information carried: fields and value types/ranges:
<Query; well formed query>

Message Book query response message

Defined Page 164
Description Response from book query: a list of book records.
Additional fields:
Coverage and overlap Full coverage, no overlap
Information carried: fields and value types/ranges:
<Query; well formed query>, <Response list; list of book records>

3.1.4 Internal Messages

Message Temporary reduction

Name Temporary reduction
Description Message to temporarily reduce a book price
Distribution Managing competition → Pricing
Purpose To temporarily reduce a book price due to a lower price elsewhere.
Information carried New price, Book ID
Coverage and overlap Full coverage, no overlap

Message Remove reduction

Name Remove reduction
Description Message indicating that temporary price reduction on a particular book should be removed.
Distribution Managing competition → Pricing
Purpose To trigger return to standard pricing for a particular book.
Information carried Book ID.
Coverage and overlap Full coverage, no overlap

Message Stock low

Name Stock low
Description Message that stock for a particular book has gone below a set threshold.
Distribution Check stock → Add to order
Purpose To trigger a process to add this item to an order, if it is not already ordered.
Information carried Book record.

Coverage and overlap Partial coverage, no overlap. No plan available if book already on order.

Message No stock

Name No stock
Description Message indicating that a book is out of stock, and is ordered.
Distribution Check stock → Out of stock response, Add to order
Purpose Trigger to order new stock outside the regular schedule.
Information carried Book ID, number ordered, date
Coverage and overlap Full coverage, Add to order overlaps with Out of stock response. Out of stock response takes precedence.

Message Get number required

Name Get number required
Description Message containing ID of book to be ordered.
Distribution Add to order → Get number by index, Get number by sales, Get number by price
Purpose To trigger calculation of number of copies to order.
Information carried Book ID
Coverage and overlap Full coverage, Get number by index overlaps with Get number by price and with Get number by sales. Get number by index has lower priority.

Message Decide supplier

Name Decide supplier
Description Message containing Book ID, number to be ordered, date and urgency.
Distribution Add to order → Decide supplier by time, Decide supplier by price
Purpose To trigger decision regarding which supplier to use.
Information carried Book record (possibly only limited fields), number required, urgency
Coverage and overlap Full coverage, Decide supplier by price overlaps with Decide supplier by time. Decide supplier by time takes precedence

Message Modify monthly order

Name Modify monthly order
Description Message to add or delete items from an order for a particular supplier.
Distribution Out of stock response → Delete items, Add to supplier order
Purpose To add or delete the indicated items from the order being developed.
Information carried Add/delete, book ID, number/all.
Coverage and overlap Full coverage, no overlap

3.1.5 Plans

Plan Respond book query

Name Respond book query
Description Queries Books and Stock DB and provides response to query.
Trigger Book query message
Context none
Incoming messages Book query message
Outgoing messages Book query response message
Used data stock DB, Books DB
Produced data List of book records
Failure Unable to access either DB
Failure recovery send message to system support. Reply with response to try again later.
Procedure:

```
Connect to Stock DB
Run query
If > 1 book record returned
    Return response
Else
    Connect to Books DB
    Run query
    Return message with response (+ not in stock annotation)
endif
```

Plan Check stock

Name Check stock
Description Determines whether book required is available in stock.none
Trigger
Context none
Incoming messages
Outgoing messages
Percepts
Actions
Used data
Produced data
Goal
Failure can't read file
Failure recovery email system support
Procedure:

```
Access stock record for book ID (from message)
If stock number > 0
```

```
   send reply "Book available"
Else
   Send reply "not in stock"
   Post message "No stock"
If reorder threshold exists
   If stock number -1 < reorder threshold
      post message stock low
   endif
Else if stock number -1 > standard reorder threshold
   Post message stock low
endif
```

Plan Add to order

Name Add to order
Description Plan to add an item to an order list, due to stock becoming low
Trigger
Context Not already ordered
Incoming messages
Outgoing messages
Percepts
Actions
Used data
Produced data
Goal
Failure Fail to write to file
Failure recovery Send email to system support. Possibly repost event, with counter to prevent infinite looping.
Procedure:

```
Subtask: Determine number to order
Subtask: Determine supplier
Subtask: Add item to monthly order for chosen supplier
```

Plan Out of stock response

Name Out of stock response
Description Places an immediate order, if appropriate, if stock is completely out and book is required.
Trigger
Context Date > 2 days from next regular order date
Incoming messages
Outgoing messages
Percepts
Actions
Used data
Produced data

Goal
Failure bounced email (Failure of action). Won't be immediately evident.
Failure recovery need general procedure for receiving, fixing and resending bounced emails.
Procedure:

```
subtask: Decide supplier (Urgent = YES)
subtask: get number required
subtask: modify monthly order, (delete items ordered now)
Action: email stock order
```

Plan Get_number_by_index

Name Get_number_by_index
Description Uses an index that classifies the type of book and provides a standard order number for that classification.
Trigger
Context none
Incoming messages
Outgoing messages
Percepts
Actions
Used data
Produced data
Goal
Failure No entry for book
Failure recovery Send message to technical support.
Procedure:

```
Identify book category as POPULAR, STANDARD, or OCCASIONAL
If category = POPULAR
   number = highorder
If category = STANDARD
   number = medium-order
If category = OCCASIONAL
   number = loworder
```

Plan get_number_by_price

Name get_number_by_price
Description Determines number of books to order dependent on cost and current cashflow.
Trigger
Context Cashflow = POOR or CRITICAL AND Price > 20
Incoming messages
Outgoing messages

Percepts
Actions
Used data
Produced data
Goal
Failure
Failure recovery
Procedure:

```
If Cashflow = CRITICAL
    Number = MIN(100/bookprice, standard-order-number)
Else IF
    Cashflow = POOR
    Number = MIN(500/bookprice, standard-order-number)
end elseif
endif
```

Plan get_number_by_sales

Name get_number_by_sales
Description Calculates number of books to order based on average monthly sales of that book.
Trigger
Context Average monthly book sales for Book ID available AND Cashflow NOT= (POOR or CRITICAL)
Incoming messages
Outgoing messages
Percepts
Actions
Used data
Produced data
Goal
Failure
Failure recovery
Procedure:

```
Number = Average monthly sales * 3
```

Plan Decide_supplier_by_time

Name Decide_supplier_by_time
Description Finds fastest reliable supplier for book
Trigger
Context Urgency = YES
Incoming messages
Outgoing messages

Percepts
Actions
Used data
Produced data
Goal
Failure No supplier found
Failure recovery Send message to manager indicating no supplier for item
Procedure:

```
Obtain record from Books DB
For each supplier in supplier list
   Obtain supplier record
   Note normal delivery time
   Note reliability index
Calculate fastest supplier
If fastest supplier reliability < "good reliability" (set values in file)
Find fastest supplier with < good reliability
If difference in delivery time (fastest, fastest reliable)
    < "small delivery difference"
   Identify and record fastest reliable supplier
Else
   Identify and record fastest supplier
```

Plan Decide_supplier_by_price

Name Decide_supplier_by_price
Description Finds cheapest supplier
Trigger
Context Urgency = NO
Incoming messages
Outgoing messages
Percepts
Actions
Used data
Produced data
Goal
Failure No supplier found
Failure recovery Email manager indicating no supplier for item
Procedure:

```
Obtain Book record
Compare book price for each supplier in supplier list
Identify and record cheapest supplier
```

Plan Add_to_supplier_order

Name Add_to_supplier_order
Description Adds item to an order from a particular supplier.

Trigger
Context supplier X
Incoming messages
Outgoing messages
Percepts
Actions
Used data
Produced data
Goal
Failure Fail to write to order file
Failure recovery Email system support. Possibly try reposting with counter to prevent infinite looping.
Procedure:

```
If order file exists for supplier X
  open order file
Else
  Create order file for supplier X
  Open file
Write items to order
Close file
```

Plan Delete_items

Name Delete_items
Description Delete items from a monthly order for a given supplier
Trigger
Context none
Incoming messages
Outgoing messages
Percepts
Actions
Used data
Produced data
Goal
Failure fail to write to file fail to open file
Failure recovery email system support
Procedure:

```
Open file for supplier X
Find item
If number required > number to delete
   modify number required to number required - number to delete
Else
   remove item
```

Plan Build monthly orders

Name Build monthly orders
Description Prepares the monthly orders and sends them out.none
Trigger
Context none
Incoming messages
Outgoing messages
Percepts
Actions
Used data
Produced data
Goal
Failure 1) Email bounces (i.e. action fails) 2) Problems reading file / corrupted file
Failure recovery 1) need general procedure for receiving, fixing and resending bounced emails. 2) Contact system support in failure method
Procedure:

```
For each supplier order file:
    Format order
    Action: Email stock order
    Annotate file with order date
    Archive file
```

3.2 Agent Sales Assistant

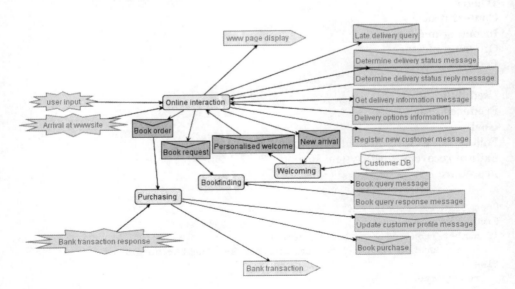

3.2.1 Capabilities

Bookfinding

Not developed.

Online interaction

Mostly not developed.

Purchasing

Not developed.

Welcoming

Not developed.

3.2.2 Processes

Mostly not developed

Process Manage Customer Profile process (SA)

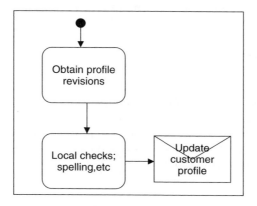

Name Manage Customer Profile process (SA)
description Obtains customer profile information, checks it locally and sends to Customer Relations agent
Activities Obtain profile revisions, Local checks
Trigger user input
Messages <Update customer profile, to Customer Relations>
Protocols Update customer profile protocol

Notes:

No further development ...

3.2.3 External Messages

Not shown

3.2.4 Internal Messages

Message New arrival

Name New arrival
Description Indicator that a new person has arrived at WWWsite.
Distribution Online interaction → Welcoming
Purpose To trigger individualized response.
Information carried Cookie/ID if available.

Message Personalized welcome

Name Personalized welcome
Description message containing individualized welcome for display to WWW page.
Distribution Welcoming → Online interaction
Purpose To provide content for display to WWW page.
Information carried Customer name, current orders, welcome text, recommendations, etc

Message Book order

Name Book order
Description The order from the WWW interface, as requested by the user.
Distribution Online interaction → Purchasing
Purpose For purchasing capability to arrange the payment and then pass onto Delivery.
Information carried Books ordered, delivery method chosen, delivery address, total cost, credit card details, customer ID

Message Book request

Name Book request
Description User request regarding a book. May include keywords, or some information fields of book record.
Distribution Online interaction → Bookfinding
Purpose To be used in formulating a well formed query for searching the Stock DB and the Books DB.
Information carried whatever user provided.

Not further developed.

3.2.5 Plans

Not developed

3.3 Agent Delivery Manager

Not developed

3.4 Agent Customer Relations

3.4.1 Capabilities

Not developed.

3.4.2 Processes

Process Manage Customer Profile process (CR)

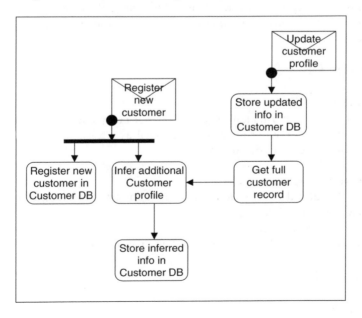

Name Manage Customer Profile process
description Process to register information about a customer when new information is provided. Includes inferring additional information.
Activities Register new customer, Store updated information, Infer additional information
Trigger Register new customer, Update customer profile
Messages none
Protocols Update customer profile protocol
Notes:

Not further developed

3.4.3 Data

Data Stock DB

Name Stock DB
Description Contains records of all books stocked
Data type Set of stock records
Included fields/aspects Each stock record contains book record, selling price, current stock number, stock order record reference, preferred suppliers.
Persistent Yes
External to system No
Initialization N/A
Produced by Stock Manager$_a$, Stock managing$_c$, Pricing$_c$, Handling new stock$_c$
Used by Stock Manager$_a$, Stock managing$_c$ Ordering$_c$, Respond book query$_p$
Used when Used when customer requests book information, book delivery goes out, new stock arrives.

Data Courier DB

Name Courier DB
Description Contains information regarding courier companies.
Data type Structure of courier records. Maintained as a readable file.
Included fields/aspects Courier records
Persistent Yes
External to system Yes
Initialization Initialized from a file (which can be updated external to the system)
Produced by Delivery Manager$_a$,
Used by Delivery Manager$_a$,
Used when Used when determining which courier to use for a delivery.

Data Postal DB

Name Postal DB

Description contains information about postal rates and approximate delivery times.
Data type Database. Use externally available DB from postal services.
Included fields/aspects not developed
Persistent Yes
External to system Yes
Initialization Build an in-memory data structure with most commonly used information.
Produced by
Used by Delivery Manager
Used when Used when providing delivery options information to customer ordering books.

Data Customer DB

Name Customer DB
Description Database of customer information, including customer profiles, purchasing history, etc. (Requires further design)
Data type Database
Included fields/aspects customer records
Persistent Yes
External to system No
Initialization N/A
Produced by Customer Relations$_a$
Used by Welcoming$_c$, Sales Assistant$_a$, Customer Relations$_a$
Used when Used to provide information for customization, marketing and any communications with customer.

Data Books DB

Name Books DB
Description Contains book records indicating potential suppliers, as well as information regarding the book.
Data type Database containing book records.
Included fields/aspects Book records
Persistent Yes
External to system No
Initialization
Produced by Cataloging$_c$, Pricing$_c$
Used by Pricing$_c$, Managing competition$_c$, Ordering$_c$, Stock managing$_c$, Respond to query$_p$, Stock Manager$a$
Used when Used to determine suppliers of books to be stocked. Used to answer user queries about books where these cannot be answered from the Stock DB.

Data Customer Orders

Name Customer Orders
Description Records of book orders placed by customers. Contains all order data, as well as data such as when it was sent, or when it is expected to be sent, etc.
Data type Set of order records.

Included fields/aspects order details, date sent, date expected, carrier, . . .
Persistent Yes
External to system No
Initialization Read in from file at start up. Written out to file as safety measure to avoid loss of data if system goes down.
Produced by Delivery Manager$_a$
Used by Delivery Manager$_a$
Used when written when order is placed. Used when order status is required or when customer complains that order is late.

Data Delivery Problems

Name Delivery Problems
Description List of records with orders which have not arrived and where the user has made a query. Includes information built up as the problem is investigated.
Data type Set of problem records.
Included fields/aspects Date contacted, order record, tracking requests, tracking responses, status, actions taken.
Persistent Yes
External to system No
Initialization N/A
Produced by Delivery Manager$_a$
Used by Delivery Manager$_a$
Used when Record started when a query is made regarding a book arrival. Used in monitoring and tracking the problem until book is found and sent to customer, or replacement book is sent. Information transferred to courier DB and to customer DB after resolution.

Data Stock Orders

Name Stock Orders
Description Records of stock orders placed
Data type set of orders to suppliers.
Included fields/aspects Order includes stock items ordered, date, expected delivery date, updates, etc.
Persistent Yes
External to system No
Initialization
Produced by Stock managing$_c$, Ordering$_c$, Delay handling$_c$, Stock manager$_a$
Used by Stock managing$_c$, Handling new stock$_c$, Stock manager$_a$
Used when Order includes stock items ordered, date, expected delivery date, updates, etc.

Data customer record

Name customer record
Description Record containing information about a customer, including contact details, profile, orders they have made, history of interactions, etc.
Data type Record structure

Included fields/aspects User supplied profile (name, address, email, phone, interests, etc.) (encrypted) credit card details, purchase history, interaction history, ...
Persistent Yes
External to system No
Initialization
Produced by Customer Relations$_a$
Used by Customer Relations$_a$
Used when updated at user request, or on interaction with user. Used for providing specific recommendations, individual welcome, etc. Also used for initializing information in customer order record.

Data customer order record

Name customer order record
Description Record of customer order. Maintained until after order (believed) delivered.
Data type Record structure
Included fields/aspects order details, date sent, date expected, carrier, ...
Persistent Yes
External to system No
Initialization Some fields initialized from customer DB where possible. (e.g. address, (partial) credit card details.
Produced by Sales assistant$_a$, Delivery Manager$_a$
Used by Delivery Manager$_a$
Used when arranging delivery. Also used if any status query, or late books query.

Data Pending Orders

Name Pending Orders
Description Records of orders which are awaiting arrival of stock.
Data type Set of order records
Included fields/aspects order details, date sent, date expected, carrier, ...
Persistent Yes
External to system No
Initialization Stored on file as safety measure. Read in from file on start up.
Produced by Ordering$_c$ Delivery Manager$_a$
Used by Delay handling$_c$, Handling new stock$_c$, Ordering$_c$, Delivery Manager$_a$
Used when used when stock arrives, to fill back-orders. Also when stock delayed to notify customer relations in case customers should be notified.

Data Monthly Order

Name Monthly Order
Description Record of books which are to be ordered in the upcoming monthly order.
Data type List of stock orders
Included fields/aspects book id, number to order, supplier, ...
Persistent Yes (only a month at a time)

External to system No
Initialization Stored on file as safety measure. Read in from file on start up.
Produced by Ordering$_c$ Delivery Manager$_a$
Used by Ordering$_c$ Delivery Manager$_a$
Used when used when system notices stock is low and adds information. Used monthly to produce regular order.

Data transaction details (temp)

Name transaction details (temp)
Description Record created for purpose of obtaining credit card payment.
Data type Transaction record
Included fields/aspects credit card number, expiry date, amount, name, card type
Persistent No
External to system No
Initialization Initialize some values from customer DB if available.
Produced by Purchasing$_c$, Sales Assistant$_a$
Used by Purchasing$_c$, Sales Assistant$_a$
Used when Used as part of book ordering process. Used for communication of transaction with bank.

Data updated info (temp)

Name updated info (temp)
Description Updated customer profile
Data type customer profile record
Included fields/aspects Name, address, email, phone, interests
Persistent No
External to system No
Initialization Initialized to current values in customer DB if these exist.
Produced by Online interaction$_c$, Sales Assistant$_a$
Used by Customer Relations$_a$
Used when Created when customer wishes to update their profile.

Data welcome text (temp)

Name welcome text (temp)
Description Created when customer wishes to update their profile.
Data type record to be converted to html page
Included fields/aspects
Persistent No
External to system No
Initialization N/A
Produced by Welcoming$_c$, Sales Assistant$_a$
Used by Online interaction$_c$, Sales Assistant$_a$
Used when Developed when user logs in.

Dictionary

Listed by name:

Name	Page	Entity
Add to order	Page 177	Plan
Add to supplier order	Page 180	Plan
Arrange delivery	Page 140	Goal
Arrival at WWWsite	Page 157	Percept
Bank transaction	Page 160	Action
Bank transaction response	Page 157	Percept
Book available	Page 166	Message
Book delivery information	Page 163	Message
Book expected soon	Page 165	Message
Book finding	Page 143	Functionality
Book finding protocol	Page 161	Protocol
Book finding scenario	Page 145	Scenario
Book located	Page 165	Message
Book order	Page 184	Message
Book ordering protocol	Page 161	Protocol
Book purchase	Page 166	Message
Book query	Page 140	Goal
Book query message	Page 164	Message
Book query response message	Page 164	Message
Book request	Page 184	Message
Book required	Page 166	Message
Bookfinding	Page 183	Capability
Books DB	Page 187	Data
Broad range of books	Page 140	Goal
Build monthly orders	Page 182	Plan
Calculate delivery time estimates	Page 140	Goal
Cataloging	Page 170	Capability
Catalogue management	Page 142	Functionality
Cheaper price notification	Page 149	Scenario
Cheaper price report	Page 158	Percept
Check stock	Page 176	Plan
Competition management	Page 144	Functionality
Competitive prices	Page 140	Goal
Confirm changes	Page 140	Goal
Courier DB	Page 186	Data
Customer DB	Page 187	Data
Customer Orders	Page 187	Data
Customer order record	Page 189	Data
Customer record	Page 188	Data
Customer Relations	Page 155	Agent
Customer contact	Page 144	Functionality

Name	Page	Entity
Monitor delivery	Page 140	Goal
Monitor stock arrivals	Page 140	Goal
New arrival	Page 184	Message
New catalogue	Page 150	Scenario
New catalogue	Page 157	Percept
No stock	Page 175	Message
No tracking response	Page 159	Percept
Not in stock	Page 166	Message
Obtain credit card details	Page 140	Goal
Obtain delivery options	Page 140	Goal
Obtain user input	Page 140	Goal
Online interaction	Page 141	Functionality
Online interaction	Page 183	Capability
Order book	Page 146	Scenario
Order status query	Page 148	Scenario
Order status querying protocol	Page 161	Protocol
Order stock	Page 140	Goal
Ordering	Page 168	Capability
Out of stock response	Page 177	Plan
Pending order arrives	Page 147	Scenario
Pending Orders	Page 189	Data
Personalized U.I.	Page 140	Goal
Personalized welcome	Page 184	Message
Place delivery request	Page 160	Action
Place order (online)	Page 140	Goal
Postal DB	Page 186	Data
Present information	Page 140	Goal
Price setting	Page 143	Functionality
Pricing	Page 171	Capability
Profile monitor	Page 142	Functionality
Provide personalized recommendations	Page 140	Goal
Provide personalized welcome	Page 140	Goal
Purchasing	Page 144	Functionality
Purchasing	Page 183	Capability
Query late books	Page 144	Scenario
Query late books protocol	Page 162	Protocol
Register customer profile	Page 140	Goal
Register new customer message	Page 166	Message
Regular order trigger	Page 159	Percept
Remove reduction	Page 174	Message
Request delivery tracking	Page 160	Action
Respond book query	Page 176	Plan
Respond to customer	Page 140	Goal
Restore book price	Page 140	Goal
Sales Assistant	Page 154	Agent

Name	Page	Entity
Send email	Page 160	Action
Set prices competitively	Page 140	Goal
Stock DB	Page 186	Data
Stock Manager	Page 156	Agent
Stock Orders	Page 188	Data
Stock arrival	Page 158	Percept
Stock arrival	Page 147	Scenario
Stock arrival delayed	Page 165	Message
Stock arrival info	Page 164	Message
Stock arrival process	Page 171	Process
Stock arrival protocol	Page 163	Protocol
Stock delay protocol	Page 163	Protocol
Stock delayed	Page 148	Scenario
Stock low	Page 174	Message
Stock maintenance process	Page 172	Process
Stock management	Page 142	Functionality
Stock managing	Page 168	Capability
Stock order	Page 148	Scenario
Stock order delay	Page 158	Percept
Temporary reduction	Page 174	Message
Tracking info	Page 157	Percept
Transaction details (temp)	Page 190	Data
Update BooksDB	Page 140	Goal
Update customer profile message	Page 164	Message
Update customer profile protocol	Page 161	Protocol
Update customer record	Page 140	Goal
Update delivery problem	Page 140	Goal
Updated info (temp)	Page 190	Data
User input	Page 156	Percept
Welcoming	Page 142	Functionality
Welcoming	Page 183	Capability
Welcome text (temp)	Page 190	Data
Worldwide sale of books	Page 140	Goal
WWW page display	Page 160	Action
WWWsite arrival	Page 149	Scenario

Listed by type:

Type	Name	Page
Action	Bank transaction	Page 160
Action	Email stock order	Page 159
Action	Place delivery request	Page 160
Action	Request delivery tracking	Page 160

Type	Name	Page
Action	Send email	Page 160
Action	WWW page display	Page 160
Agent	Customer Relations	Page 155
Agent	Delivery Manager	Page 155
Agent	Sales Assistant	Page 154
Agent	Stock Manager	Page 156
Capability	Bookfinding	Page 183
Capability	Cataloging	Page 170
Capability	Delay handling	Page 169
Capability	Handling new stock	Page 169
Capability	Managing competition	Page 170
Capability	Online interaction	Page 183
Capability	Ordering	Page 168
Capability	Pricing	Page 171
Capability	Purchasing	Page 183
Capability	Stock managing	Page 168
Capability	Welcoming	Page 183
Data	Books DB	Page 187
Data	Courier DB	Page 186
Data	Customer DB	Page 187
Data	Customer Orders	Page 187
Data	Customer order record	Page 189
Data	Customer record	Page 188
Data	Delivery Problems	Page 188
Data	Pending Orders	Page 189
Data	Postal DB	Page 186
Data	Stock DB	Page 186
Data	Stock Orders	Page 188
Data	Transaction details (temp)	Page 190
Data	Updated info (temp)	Page 190
Data	Welcome text (temp)	Page 190
Functionality	Book finding	Page 143
Functionality	Catalogue management	Page 142
Functionality	Competition management	Page 144
Functionality	Customer contact	Page 144
Functionality	Delivery handling	Page 143
Functionality	Lost goods management	Page 143
Functionality	Online interaction	Page 141
Functionality	Price setting	Page 143
Functionality	Profile monitor	Page 142
Functionality	Purchasing	Page 144
Functionality	Stock management	Page 142
Functionality	Welcoming	Page 142
Goal	Arrange delivery	Page 140
Goal	Book query	Page 140

Type	Name	Page
Goal	Broad range of books	Page 140
Goal	Calculate delivery time estimates	Page 140
Goal	Competitive prices	Page 140
Goal	Confirm changes	Page 140
Goal	Delivery of books	Page 140
Goal	Delivery tracking	Page 140
Goal	Determine delivery status	Page 140
Goal	Fast, reliable service	Page 140
Goal	Fill pending order	Page 140
Goal	Fully online system	Page 140
Goal	Have books in stock	Page 140
Goal	Identify affected orders	Page 140
Goal	Inform customer	Page 140
Goal	Locating of books	Page 140
Goal	Log books arriving	Page 140
Goal	Log books outgoing	Page 140
Goal	Log delivery problems	Page 140
Goal	Log outgoing delivery	Page 140
Goal	Log tracking information	Page 140
Goal	Lower book price	Page 140
Goal	Make payment (online)	Page 140
Goal	Manage new stock	Page 140
Goal	Monitor competitive response	Page 140
Goal	Monitor delivery	Page 140
Goal	Monitor stock arrivals	Page 140
Goal	Obtain credit card details	Page 140
Goal	Obtain delivery options	Page 140
Goal	Obtain user input	Page 140
Goal	Order stock	Page 140
Goal	Personalized U.I.	Page 140
Goal	Place order (online)	Page 140
Goal	Present information	Page 140
Goal	Provide personalized recommendations	Page 140
Goal	Provide personalized welcome	Page 140
Goal	Register customer profile	Page 140
Goal	Respond to customer	Page 140
Goal	Restore book price	Page 140
Goal	Set prices competitively	Page 140
Goal	Update BooksDB	Page 140
Goal	Update customer record	Page 140
Goal	Update delivery problem	Page 140
Goal	Worldwide sale of books	Page 140
Message	Book available	Page 166
Message	Book delivery information	Page 163
Message	Book expected soon	Page 165

Type	Name	Page
Message	Book located	Page 165
Message	Book order	Page 184
Message	Book purchase	Page 166
Message	Book query message	Page 164
Message	Book query response message	Page 164
Message	Book request	Page 184
Message	Book required	Page 166
Message	Decide supplier	Page 175
Message	Delivery options information	Page 165
Message	Determine delivery status message	Page 163
Message	Determine delivery status reply message	Page 164
Message	Get delivery information message	Page 165
Message	Get number required	Page 175
Message	Investigating	Page 165
Message	Late delivery query	Page 163
Message	Modify monthly order	Page 175
Message	New arrival	Page 184
Message	No stock	Page 175
Message	Not in stock	Page 166
Message	Personalized welcome	Page 184
Message	Register new customer message	Page 166
Message	Remove reduction	Page 174
Message	Stock arrival delayed	Page 165
Message	Stock arrival info	Page 164
Message	Stock low	Page 174
Message	Temporary reduction	Page 174
Message	Update customer profile message	Page 164
Percept	Arrival at WWWsite	Page 157
Percept	Bank transaction response	Page 157
Percept	Cheaper price report	Page 158
Percept	Failed stock arrival	Page 158
Percept	New catalogue	Page 157
Percept	No tracking response	Page 159
Percept	Regular order trigger	Page 159
Percept	Stock arrival	Page 158
Percept	Stock order delay	Page 158
Percept	Tracking info	Page 157
Percept	User input	Page 156
Plan	Add to order	Page 177
Plan	Add to supplier order	Page 180
Plan	Build monthly orders	Page 182
Plan	Check stock	Page 176
Plan	Decide supplier by price	Page 180
Plan	Decide supplier by time	Page 179
Plan	Delete items	Page 181

B

Descriptor Forms

This appendix contains each of the descriptors, with all their fields. Templates can be downloaded from *http://www.cs.rmit.edu.au/agents/prometheus*. The web site also contains printable versions, which are suitable for printing out and using in classes or workshops for paper-based design work. An alternative to using paper or electronic templates is to use the Prometheus Design Tool (PDT) that can be freely downloaded from *http://www.cs.rmit.edu.au/agents/pdt*.

Many of the fields in these forms are redundant with respect to the diagrams. For example, given the system overview diagram, one can determine for each agent what percepts it handles, what actions it performs, what messages it sends to other agents and which agents these messages are sent to and what external data it accesses. By using the Prometheus Design Tool, the contents of such fields can be automatically derived.

Developing Intelligent Agent Systems   L. Padgham & M. Winikoff
© 2004 John Wiley & Sons, Ltd   ISBN: 0-470-86120-7 (HB)

Goal Descriptor

Name:
Description:
Subgoals:

Functionality Descriptor

Name:
Description:
Triggers:
Actions:
Information Used:
Information Produced:
Goals:

Agent Descriptor

Name:
Description:
Cardinality: (min-max)
Lifetime:
Initialisation:
Demise:
Incoming and Outgoing Messages: For each indicate the source and destination,
e.g. acknowledge (AnAgent $\rightarrow$ AnotherAgent)
Internal Messages: For each indicate the source and destination,
e.g. CheckAvailability (MyPlan $\rightarrow$ MyCapability)
Percepts:
Actions:
Uses data:
Produces data:
Internal data:
Goals:
Functionalities:
Protocols:
Included plans:
Included capabilities:

Capability Descriptor

Name:
Description:
Goals:
Processes:
Protocols:
Incoming and Outgoing Messages: For each indicate the source and destination,
e.g. acknowledge (ACapability $\rightarrow$ AnotherCapability)
Internal Messages: For each indicate the source and destination,
e.g. CheckAvailability (MyPlan $\rightarrow$ MyCapability)
Percepts:
Actions:
Uses data:
Produces data:
Internal data:
Included plans:
Included capabilities:
Notes:

Plan Descriptor

Name:
Description:
Trigger:
Context:
Incoming and Outgoing Messages: For each indicate the source and destination,
e.g. acknowledge (APlan$_p$ $\rightarrow$ AgentX$_a$) message2 (APlan$_p$ $\rightarrow$ AnotherPlan$_p$)
Percepts:
Actions:
Uses data:
Produces data:
Goal:
Failure:
Failure Recovery:
Procedure:

Percept Descriptor

Name:
Description:
Information Carried:
Knowledge Updated:
Source:
Processing:
Agents responding:
Expected Frequency:

Action Descriptor

Name:
Description:
Parameters:
Duration:
Failure:
Partial Change:
Side Effects:

Message Descriptor

Name:
Description:
Distribution: List of *Sender* → *Receiver* pairs.
Purpose:
Carried Information:
Coverage & Overlap: A message is *covered* if there will always be at least one applicable plan to handle it; otherwise it is uncovered. A message has no overlap if there is always at most one applicable plan to handle it; otherwise it has overlap.

Data Descriptor

Name:
Description:
Data type:
Included fields/aspects:
Persistent: (yes/no)
External to system: (yes/no)
Produced by:
Used by:
Used when:

Protocol Descriptor

Name:
Description:
Included Messages: For each indicate the source and destination, e.g. request (AnAgent →
AnotherAgent)
Scenarios:
Agents:
Notes:

Scenario Descriptor

Name:
Description:
Trigger:
Steps: Type is one of ACTION, PERCEPT, GOAL, SCENARIO or OTHER

#	Type	Name	Functionality	Description	Data produced	Data used

Variations:

Process Descriptor

Name:
Description:
Triggers:
Activities:
Messages: For each message type give the destination
Protocols:
Capabilities:

C

The AUML Notation

This appendix provides a brief introduction to those parts of the AUML-2 notation which we use with Prometheus. For more details, see the AUML website (www.auml.org). In the following description, 'AUML' refers to the revised version of AUML. We will use 'AUML-1' to explicitly refer to the older version of AUML and will use 'AUML-2' to explicitly refer to the revised version of AUML where confusion may otherwise arise.

This description of AUML-2 is necessarily out of date – AUML-2 is still evolving. We have tried to reduce the likely impact of changes by using a subset of AUML that we believe is likely to remain stable (and that is very similar to UML 2.0 (OMG 2003)).

☞ NOTE: In order to understand the relationships between AUML-1, AUML-2 and UML 2.0, it can be useful to have some idea of the history. At the time that AUML-1 was developed, UML provided sequence diagrams that captured a single interaction scenario. AUML's contribution was to provide a notation that allowed all possible interaction scenarios to be captured in a single diagram. UML version 2.0 was developed after AUML-1. UML 2.0 extended sequence diagrams so they could either be *instance* (i.e. capture a single interaction scenario) or *generic* (i.e. cover a number of possible interaction scenarios). AUML-2 was developed after UML 2.0 and in many ways is closer to UML 2.0 than it is to AUML-1. However, there are differences between UML 2.0 and AUML-2, and AUML-2 is (at the time of writing) still being developed.

Developing Intelligent Agent Systems L. Padgham & M. Winikoff
© 2004 John Wiley & Sons, Ltd ISBN: 0-470-86120-7 (HB)

The AUML notation is an extension of interaction diagram. As such, an AUML sequence diagram[1] has lifelines for each agent with messages depicted by arrows between the lifelines and with time increasing as one moves downward. So, the example below shows a User agent sending a Query message to a System agent followed by the System agent sending a Response message to the User agent.

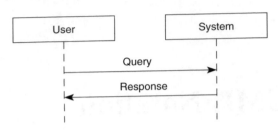

AUML places this within a frame. The 'sd' stands for 'sequence diagram' and is followed by the name of the sequence diagram.

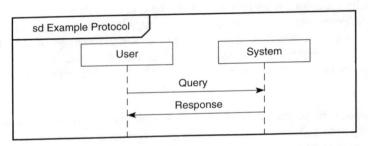

The primary way that AUML allows alternatives, parallelism, and so on to be specified is using *boxes*. A box is a region within the sequence diagram that contains messages and may contain nested boxes. Each box has a tag that describes the type of box (e.g. Alternative, Parallel, Option, etc.). A box can affect the interpretation of its contents in a range of ways depending on its type. For example, an Option box indicates that its contents may be executed as normal, or may be not executed at all (in which case the interpretation of the sequence diagram continues after the Option box). Whether the box is executed can be specified by *guards*. A guard, denoted by text in square brackets, indicates a condition that must be true in order for the Option box to execute.

The example below uses an Option box and says that after the User sends their Query to the System, the System may reply with a Response (if the system is operational, specified in the guard). If the system is not operational, then nothing happens in response to the user's Query.

---

[1] AUML-2 changed terminology from 'interaction diagram' (used in AUML-1) to 'sequence diagram' in order to be consistent with UML 2.0.

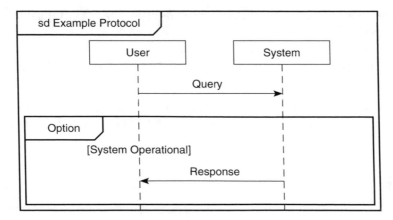

Most box types can be divided into *regions* indicated by heavy horizontal dashed lines. For example, an Alternative box can have a number of regions (each with its own guard) and exactly one region will be executed. The example below shows an example of nested boxes. The Option box, as before, indicates that nothing happens if the system is not operational. If the system is operational, then we have two Alternatives (separated by a horizontal heavy dashed line). The first alternative is that the System sends the user a Response message. The second alternative is that the System indicates that the user's query was not understood. In this example, there are no guards, so the sequence diagram indicates that the System can respond with two different messages, but does not indicate under what conditions a given response is chosen.

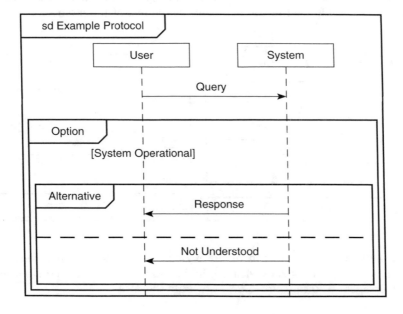

The following are some of the box types that are defined by AUML (and include all of the box types that we use):

- **Alternative:** Specifies that one of the box's regions occurs. One of the regions may have 'else' as the guard.

- **Option:** Can only have a single region. Specifies that this region may or may not occur. An Option box is equivalent to an Alternative tag with an empty second region:

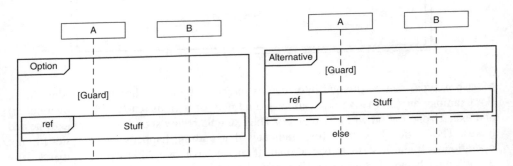

- **Break:** Either none of the regions of the box occur (in which case the box is treated as if it were absent) or one of the regions occurs. If one of the regions occurs, then the end of the Break box terminates the protocol. A Break box is equivalent to an Alternative where the Alternative box contains the contents of the Break box and also has an additional region that contains the rest of the protocol.

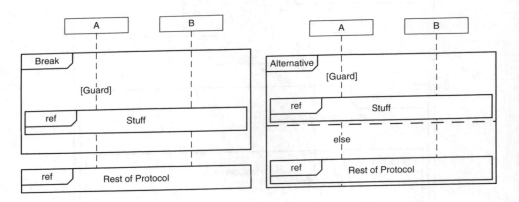

- **Parallel:** Specifies that each of the regions takes place simultaneously and the sequence of messages is interleaved. In the example below, the possible sequences of messages are qrst, qsrt, qstr, sqrt, sqtr and stqr.

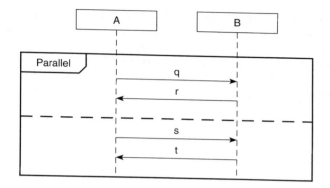

- **Critical Region:** Specifies that no interleaving should take place with the contents of the box. In the example below, messages q or r cannot occur between s and t so the possible sequences of messages are qrst, qstr, and stqr.

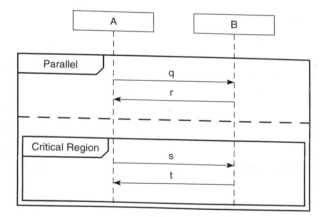

- **Loop:** Can only have a single region. Specifies that the region is repeated some number of times. The tag gives the type ('Loop') and also an indication of the number of repetitions that can be a fixed number (or a range) or a Boolean condition. The current AUML specification is not very clear on the format, so the examples below are based on UML 2.0 (OMG 2003, page 413).

  - Loop(1,3) – at least one repetition, at most 3 repetitions
  - Loop(1,*) – at least one repetition, no upper limit ('*' specifies infinity)
  - Loop – same as Loop(0,*), i.e. any number of repetitions
  - Loop(2,2) – exactly two repetitions
  - Loop(3) – same as Loop(3,3)

- **Ref:** This box type is a little different in that it does not contain sub-boxes or messages. Instead it contains the name of another protocol. This is basically a form of procedure call – the interpretation of the Ref box is obtained by replacing it with the protocol it refers to. The example below shows an existing authentication protocol being used before the user issues a query. The name of the protocol is given in the body of the box and lifelines are not shown within the Ref box.

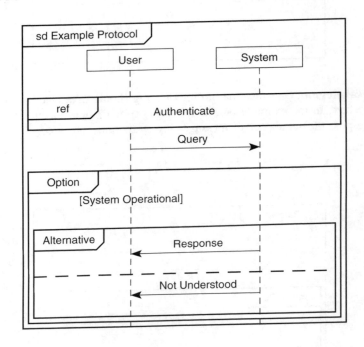

Although boxes are the primary mechanism that AUML provides for describing control flow, there is also a goto mechanism, 'continuations'. There are two types of continuations: incoming (labels) and outgoing (gotos). When interaction reaches an outgoing continuation, it continues at the incoming continuation with the same name. Each outgoing continuation must have exactly one matching incoming continuation. Both continuations are depicted by rounded rectangles; outgoing continuations (goto) have a right-pointing triangle on their right side, whereas incoming continuations (labels) have a right-pointing triangle on their left side (see Figures C.1 and C.2 for examples).

The example in Figure C.1 shows the use of continuations to specify a repeated interaction: if the system's response to the user is either Response or Not Understood, then interaction continues with the user sending another Query.

Continuations are gotos, and although they can be useful for describing exceptional conditions or for terminating an interaction, they should be used with care. Overuse of continuations yields sequence diagrams that are difficult to read, understand and modify.

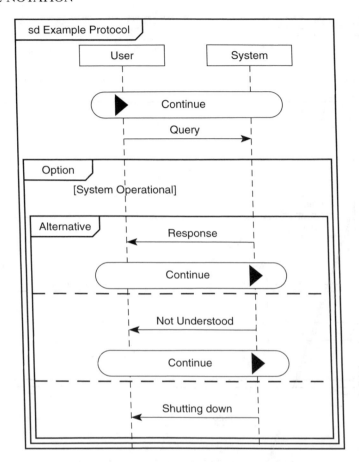

Figure C.1 Example of AUML continuations

The example in Figure C.2 demonstrates the features of AUML that we have discussed, including all of the features that we use. The figure shows the top-level interaction of a hypothetical database system with a backup that is kept up to date. The top-level interaction is a loop. Within this loop, there are three alternatives:

1. The user can send a Retrieve message and receive a Results.

2. The user can specify an Update to the data. After this message is received by the system, two things happen in parallel:

   • The system responds to the user (but only if tracing has been enabled).

   • The system updates the backup. The Critical Region box specifies that nothing else should occur between the Update request to the Backup and the Backup's response.

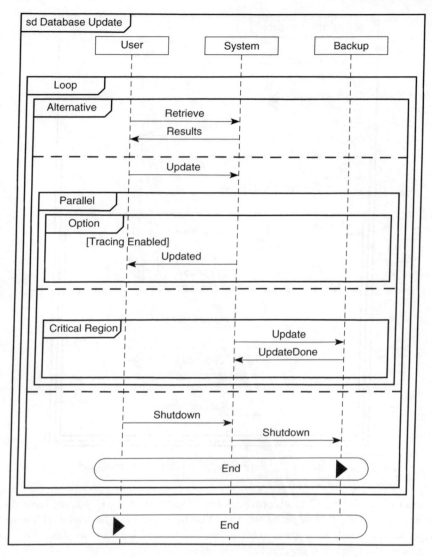

Figure C.2   AUML example showing notational elements

3. The user can ask the system to shutdown. The system relays the request to the backup and then the interaction jumps to the end.

In addition to sequence diagrams, AUML (and UML 2.0) also provides *Interaction Overview Diagrams*. These extend activity diagrams by allowing sequence diagrams to appear as building blocks. They tend to become very cluttered if the sequence diagrams are too fine grained (e.g. individual messages) and so they are useful more as a means of structuring protocols than as a means of specifying protocols. For more details, see (Huget *et al.* 2003).

☞ NOTE: There are minor differences between AUML and UML 2.0. For example, the notation for continuations in AUML has a triangle indicating whether it is incoming (label) or outgoing (goto). In AUML, the names of boxes are used, whereas UML uses abbreviations (e.g. 'Alternative' becomes 'alt').

Bibliography

AAII 1996 dMARS Technical Overview. *The dMARS V1.6.11 System Overview.*

Booch G, Rumbaugh J and Jacobson I 1999 *The Unified Modeling Language User Guide*, Object Technology Series. Addison-Wesley.

Bratman ME 1987 *Intentions, Plans, and Practical Reason.* Harvard University Press, Cambridge, MA.

Brazier FMT, Dunin-Keplicz BM, Jennings NR and Treur J 1997 DESIRE: Modelling multi-agent systems in a compositional formal framework. *International Journal of Cooperative Information Systems* **6**(1), 67–94.

Bresciani P, Giorgini P, Giunchiglia F, Mylopoulos J and Perini A 2002 Tropos: An agent-oriented software development methodology. Technical Report DIT-02-0015, University of Trento, Department of Information and Communication Technology, Trento, Italy.

Burmeister B 1996 Models and methodology for agent-oriented analysis and design. *Working Notes of the KI'96 Workshop on Agent Oriented Programming and Distributed Systems.*

Burrafato P and Cossentino M 2002 Designing a multi-agent solution for a bookstore with the PASSI methodology. *Proceedings of the Fourth International Bi-Conference Workshop on Agent-Oriented Information Systems (AOIS-2002)*, Toronto, Canada. Available from *http://mozart.csai.unipa.it/passi/*.

Busetta P, Rönnquist R, Hodgson A and Lucas A 1999 JACK – Components for Intelligent Agents in Java. Technical Report No. 1, Agent Oriented Software Pty. Ltd, Melbourne, Australia. Available from *http://www.agent-software.com*.

Bush G, Cranefield S and Purvis M 2001 The Styx agent methodology. The Information Science Discussion Paper Series 2001/02, Department of Information Science, University of Otago, Otago, New Zealand. Available from *http://divcom.otago.ac.nz/infosci*.

Caire G, Leal F, Chainho P, Evans R, Garijo F, Gomez J, Pavon J, Kearney P, Stark J and Massonet P 2002 Agent oriented analysis using MESSAGE/UML. In *Agent-Oriented Software Engineering II Second Internation Workshop, AOSE 2001*, Montreal, Canada. Published as Lecture Notes in Computer Science, Vol. 2222 (eds. Wooldridge M, Ciancarini P and Weiss G), pp. 101–108. Springer.

Cernuzzi L and Rossi G 2002 On the evaluation of agent oriented modeling methods. *Proceedings of the OOPSLA 2002 Workshop on Agent-Oriented Methodologies*, Seattle, WA. Published by the Centre for Object Technology Applications and Research (COTAR) at the University of Technology, Sydney (eds. Debenham J, Henderson-Sellers B, Jennings N and Odell J).

Cheyer A and Martin D 2001 The open agent architecture. *Journal of Autonomous Agents and Multi-Agent Systems* **4**(1), 143–148.

Cohen PR and Levesque HJ 1991 Teamwork. *Nous* **25**(4), 487–512.

Developing Intelligent Agent Systems   L. Padgham & M. Winikoff
© 2004 John Wiley & Sons, Ltd   ISBN: 0-470-86120-7 (HB)

Collinot A, Drogoul A and Benhamou P 1996 Agent oriented design of a soccer robot team. *Proceedings of ICMAS'96, Kyoto, Japan*.

Cossentino M and Potts C 2002 A CASE tool supported methodology for the design of multi-agent systems. *Proceedings of the International Conference on Software Engineering Research and Practice (SERP'02)*, Las Vegas. Available from *http://mozart.csai.unipa.it/passi/*.

Cost RS, Chen Y, Finin T, Labrou Y and Peng Y 1999 Using colored Petri nets for conversation modeling. *Workshop on Agent Communication Languages at the Sixteenth International Joint Conference on Artificial Intelligence (IJCAI-99)*, Stockholm, Sweden. Available from *http://www.csee.umbc.edu/~jklabrou/*.

Dam KH 2003 Evaluating agent-oriented software engineering methodologies. Master's Thesis, School of Computer Science and Information Technology, RMIT University, Melbourne, Australia. (supervisors: Michael Winikoff and Lin Padgham).

Dam KH and Winikoff M 2003 Comparing agent-oriented methodologies. In *Proceedings of the Fifth International Bi-Conference Workshop on Agent-Oriented Information Systems* (ed. Giorgini P, Henderson-Sellers B and Winikoff M), pp. 52–59. Melbourne, Australia.

Debenham J and Henderson-Sellers B 2002 Full lifecycle methodologies for agent-oriented systems – the extended OPEN process framework. *Proceedings of Agent-Oriented Information Systems (AOIS-2002) at CAiSE'02*, Toronto, Canada.

DeLoach SA 2001 Analysis and design using MaSE and agentTool. *Proceedings of the 12th Midwest Artificial Intelligence and Cognitive Science Conference (MAICS 2001)*.

DeLoach SA, Wood MF and Sparkman CH 2001 Multiagent systems engineering. *International Journal of Software Engineering and Knowledge Engineering* **11**(3), 231–258.

Dennett DC 1987 *The Intentional Stance*. MIT Press.

Drogoul A and Zucker J 1998 Methodological issues for designing multi-agent systems with machine learning techniques: capitalizing experiences from the robocup challenge. Technical Report LIP6 1998/041, Laboratoire d'Informatique de Paris 6.

Elammari M and Lalonde W 1999 An agent-oriented methodology: high-level and intermediate models. In Proceedings of the 1st International Workshop on Agent-Oriented Information Systems (eds. Wagner G and Yu E) Available from *http://www.aois.org*.

Fowler M and Kendall 2003 *UML Distilled: A Brief Guide to the Standard Object Modeling Language*, third edition, Object Technology Series. Addison-Wesley.

Georgeff MP and Lansky AL 1986 Procedural knowledge. *Proceedings of the IEEE* Special Issue on Knowledge Representation **74**, 1383–1398.

Georgeff M and Rao A 1998 Rational software agents: from theory to practice. In *Agent Technology: Foundations, Applications, and Markets* (eds. Jennings NR and Wooldridge MJ), Chapter 8, pp. 139-160, Springer.

Georgeff M, Pell B, Pollack M, Tambe M and Wooldridge M 1999 The belief-desire-intention model of agency. *Proceedings of Agent Theories, Architectures, and Languages ATAL '98*, 1998. Published as Lecture Notes in Computer Science, Vol. 1555 (eds. Giunchiglia F, Odell J and Weiss G). Springer, 1999.

Giunchiglia F, Mylopoulos J and Perini A 2002 The Tropos software development methodology: processes, models and diagrams. *Third International Workshop on Agent-Oriented Software Engineering* Lecture Notes in Computer Science series (LNCS 2585), (eds. Giunchiglia F, Odell J and Weiß G), pp. 162–173. Springer.

Glaser N 1996 The CoMoMAS methodology and environment for multi-agent system development. In *Multi-Agent Systems Methodologies and Applications* (eds. Zhang C and Lukose D), pp. 1-16. Springer LNAI 1286. *Second Australian Workshop on Distributed Artificial Intelligence*.

Harrison CG, Chess DM and Kershenbaum A 1995 Mobile Agents: Are they a good idea? Technical Report No. RC 19887, T. J. Watson Research Center, Yorktown Heights, New York.

Hendler J 2001 Agents and the semantic web. *IEEE Intelligent Systems* **16**(2), 30–37.

Huber MJ 1999 Jam: A BDI-theoretic mobile agent architecture. *Proceedings of the Third International Conference on Autonomous Agents, (Agents'99)*, pp. 236-243. Seattle, WA.

Huget MP 2002 Nemo: an agent-oriented software engineering methodology. *Proceedings of the OOPSLA 2002 Workshop on Agent-Oriented Methodologies*, Seattle, WA. Published by the Centre for Object Technology, Sydney (eds. Debenham J, Henderson-Sellers B, Jennings N and Odell J).

Huget MP, Odell J, Haugen Ø, Nodine MM, Cranefield S, Levy R and Padgham. L 2003 FIPA modeling: interaction diagrams on *www.auml.org* under "Working Documents". FIPA Working Draft (version 2003-07-02).

Iglesias C, Garijo M and González J 1999 A survey of agent-oriented methodologies. In *ATAL-98* (ed. Müller J, Singh MP and Rao AS), pp. 317–330. Springer-Verlag, Heidelberg, Germany.

Iglesias CA, Garijo M, González JC and Velasco JR 1997 Analysis and design of multiagent systems using MAS-commonKADS. *Proceedings of Agent Theories, Architectures, and Languages (ATAL)*, 1997. Published as Lecture Notes in Computer Science, Vol. 1365 (eds. Singh MP, Rao AS and Wooldridge M). Springer, 1998.

Ingrand FF, Georgeff MP and Rao AS 1992 An architecture for real-time reasoning and system control. *IEEE Expert* **7**(6), 34–44.

Jennings NR 2001 An agent-based approach for building complex software systems. *Communications of the ACM* **44**(4), 35–41.

Jennings NR and Wooldridge MJ (ed.) 1998a *Agent Technology: Foundations, Applications, and Markets*. Springer.

Jennings N and Wooldridge M 1998b Applications of intelligent agents. In *Agent Technology: Foundations, Applications, and Markets* (eds. Jennings NR and Wooldridge MJ), Chapter 1, pp. 3-28. Springer.

Jennings NR, Faratin P, Norman TJ, O'Brien P and Odgers B 2000a Autonomous agents for business process management. *International Journal of Applied Artificial Intelligence* **14**(2), 145–189.

Jennings NR, Faratin P, Norman TJ, O'Brien P, Odgers B and Alty JL 2000b Implementing a business process management system using ADEPT: A real-world case study. *International Journal of Applied Artificial Intelligence* **14**(5), 421–465.

Juan T, Pearce A and Sterling L 2002 ROADMAP: Extending the Gaia methodology for complex open systems. *Proceedings of the First International Joint Conference on Autonomous Agents and Multi-Agent Systems (AAMAS 2002)*, Bologna, Italy, pp. 3–10. ACM Press.

Kendall EA, Malkoun MT and Jiang CH 1996 A methodology for developing agent based systems. In *Distributed Artificial Intelligence: Architecture and Modelling, First Australian Workshop on DAI*, Canberra, ACT, Australia, November 13, 1995. Published as Lecture Notes in Computer Science. Vol. 1087 (ed. Zhang C and Lukose D). Springer.

Kinny D and Georgeff M 1996 Modelling and design of multi-agent systems. *Agent Theories, Architectures, and Languages (ATAL)*, 1996. Published as Lecture Notes in Computer Science, Vol. 1193 (eds. Müller JP, Wooldridge M and Jennings NE). Springer, 1997.

Kinny D, Georgeff M and Rao A 1996 A methodology and modelling technique for systems of BDI agents. *Agents Breaking Away, 7th European Workshop on Modelling Autonomous Agents in a Multi-Agent World*, Eindhovenm, The Netherlands, January 22–25, 1996, Proceedings. Lecture Notes in Computer Science 1038 (eds, Van de Velde W and Perram W), pp. 56–71. Springer.

Kotz D and Gray B 1999 Mobile agents and the future of the Internet. *ACM Operating Systems Review* **33**(3), 7–13.

Kruchten P 1998 *The Rational Unified Process*, Object Technology Series. Addison-Wesley.

Lee J, Huber MJ, Kenny PG and Durfee EH 1994 UM-PRS: An implementation of the procedural reasoning system for multirobot applications. *Proceedings of the Conference on Intelligent Robotics in Field, Factory, Service, and Space (CIRFFSS'94)*, pp. 842–849.

Lind J 2000 A development method for multiagent systems. *Cybernetics and Systems: Proceedings of the 15th European Meeting on Cybernetics and Systems Research, Symposium "From Agent Theory to Agent Implementation"*.

Liu L and Yu E 2001 From requirements to architectural design - using goals and scenarios *ICSE-2001 Workshop: From Software Requirements to Architectures (STRAW 2001)*, Toronto, Canada, pp. 22–30.

Luck M, Ashri R and d'Inverno M 2004 *Agent-Based Software Development*. Artech House. ISBN 1-58053-605-0.

Luck M, McBurney P and Preist C 2003 *Agent Technology: Enabling Next Generation Computing (A Roadmap for Agent Based Computing)*, AgentLink. ISBN 0854 327886. Available from *www.agentlink.org/roadmap*.

Maes P 1994 Agents that reduce work and information overload. *Communications of the ACM* **37**(7), 31–40.

Mathieu P, Routier JC and Secq Y 2003 Towards a pragmatic methodology for open multi-agent systems. *Foundations of Intelligent Systems, 14th International Symposium*, ISMIS 2003, Maebashi City, Japan, October 28–31, 2003, Proceedings. Lecture Notes in Computer Science 2871 (eds. Zhong N, Ras ZW, Tsumoto S and Suzuki E), pp. 206–210. Springer.

McIlraith SA, Son TC and Zeng H 2001 Semantic web services. *IEEE Intelligent Systems* **16**(2), 46–53.

Moreau L 2002 Agents for the grid: a comparison for Web services (Part 1: the transport layer). In *Second IEEE/ACM International Symposium on Cluster Computing and the Grid (CCGRID 2002)* (eds. Bal HE, Lohr KP and Reinefeld A), pp. 220–228. IEEE Computer Society, Berlin, Germany.

Moreau L, Miles S, Goble C, Greenwood M, Dialani V, Addis M, Alpdemir N, Cawley R, Roure DD, Ferris J, Gaizauskas R, Glover K, Greenhalgh C, Li P, Liu X, Lord P, Luck M, Marvin D, Oinn T, Paton N, Pettifer S, Radenkovic MV, Roberts A, Robinson A, Rodden T, Senger M, Sharman N, Stevens R, Warboys B, Watson P and Wroe C 2002 On the Use of Agents in a BioInformatics Grid. *Network Tools and Applications in Biology (NETTAB'2002) — Agents in Bioinformatics*, Bologna, Italy.

Muscettola N, Nayak PP, Pell B and Williams B 1998 Remote agent: To boldly go where no AI system has gone before. *Artificial Intelligence* **103**(1-2), 5–48.

Nowostawski M, Purvis M and Cranefield S 2001 A layered approach for modelling agent conversations. *Proceedings of the 2nd International Workshop on Infrastructure for Agents, MAS, and Scalable MAS, 5th International Conference on Autonomous Agents*, pp. 163-170. Montreal, Canada.

Odell J, Parunak H and Bauer B 2000 Extending UML for agents. *Proceedings of the Agent-Oriented Information Systems Workshop at the 17th National conference on Artificial Intelligence*, pp. 3–17.

Odell J 2002 Objects and agents compared. *Journal of Object Technology* **1**(1), 41–53.

O'Malley SA and DeLoach SA 2001 Determining when to use an agent-oriented software engineering. *Proceedings of the Second International Workshop On Agent-Oriented Software Engineering (AOSE-2001)*, pp. 188–205. Montreal, Canada.

OMG 2003 UML 2.0 superstructure specification object management group. Available from *www.omg.org*, document ptc/03-08-02.

Parunak HVD 1997 "go to the ant": Engineering principles from natural multi-agent systems. *Annals of Operations Research* **75**, 69–101. (Special Issue on Artificial Intelligence and Management Science).

Poutakidis D, Padgham L and Winikoff M 2002 Debugging multi-agent systems using design artifacts: the case of interaction protocols. *Proceedings of the First International Joint Conference on Autonomous Agents and Multi Agent Systems (AAMAS'02)*, Bologna, Italy pp. 960–967.

Poutakidis D, Padgham L and Winikoff M 2003 An exploration of bugs and debugging in multi-agent systems. *Proceedings of the 14th International Symposium on Methodologies for Intelligent Systems (ISMIS)*, pp. 628–632. Maebashi City, Japan.

Rao AS and Georgeff MP 1991 Modeling rational agents within a BDI-Architecture. In *Procedings of the 2nd International Conference on Priniciples of Knowledge Representation and Reasoning (KR'91)*. Cambridge, MA, USA, April 22–25, 1991 (eds. Allen J, Fikes R and Sandewall E), pp. 473–484 Morgan Kaufmann.

Rao AS and Georgeff MP 1992 An abstract architecture for rational agents. In *Proceedings of the 3rd International Conference on Principles of Knowledge Representation and Reasoning*, Cambridge, MA October 25–29 1992 (eds. Rich C, Swartout W and Nebel B), pp. 439–449. Morgan Kaufmann Publishers, San Mateo, CA.

Reisig W 1985 *Petri Nets: An Introduction*, EATCS Monographs on Theoretical Computer Science. Springer-Verlag. ISBN 0-387-13723-8.

Rolland C, Grosz G and Kla R 1999 Experience with goal-scenario coupling in requirements engineering. *Proceedings of the Fourth IEEE International Symposium on Requirements Engineering (RE'99)* Limerick, Ireland.

Russell S and Norvig P 1995 *Artificial Intelligence: A Modern Approach*. Prentice Hall.

Shehory O and Sturm A 2001 Evaluation of modeling techniques for agent-based systems. *Proceedings of the Fifth International Conference on Autonomous Agents*, Montreal, Canada pp. 624–631. ACM Press.

Shen W and Norrie D 1999 Agent-based systems for intelligent manufacturing: a state-of-the-art survey. *Knowledge and Information Systems, An International Journal* 1(2), 129–156. Extended version available online at *http://imsg.enme.ucalgary.ca/publication/abm.htm*.

Sturm A and Shehory O 2002 Towards industrially applicable modeling technique for agent-based systems (poster). *Proceedings of International Conference on Autonomous Agents and Multi-Agent Systems* (AAMAS 2002), Bologna. ACM Press.

Sturm A and Shehory O 2003 A framework for evaluating agent-oriented methodologies. In *Proceedings of the Fifth International Bi-Conference Workshop on Agent-Oriented Information Systems* (eds. Giorgini P and Winikoff M), pp. 60–67. Melbourne, Australia.

Thangarajah J, Padgham L and Harland J 2002 Representation and reasoning for goals in BDI agents. *Australasian Computer Science Conference*, Jan 2002.

Tidhar G, Heinze C and Selvestrel M 1998 Flying together: modelling air mission teams. *Applied Intelligence* 8(3), 195–218.

van Lamsweerde A 2001 Goal-oriented requirements engineering: a guided tour. *Proceedings of the 5th IEEE International Symposium on Requirements Engineering (RE'01)*, pp. 249–263, Toronto, Canada.

Varga LZ, Jennings NR and Cockburn D 1994 Integrating intelligent systems into a cooperating community for electricity distribution management. *Int Journal of Expert Systems with Applications* 7(4), 563–579.

Wagner G 2002 A UML profile for external AOR models. In *Proceedings of Third International Workshop on Agent-Oriented Software Engineering (AOSE-2002)*, held at *Autonomous Agents & Multi-Agent Systems (AAMAS 2002)*, Palazzo Re Enzo, Bologna, Italy, July 15, 2002, Springer-Verlag LNAI 2585.

Wagner G 2003 The agent-object-relationship metamodel: towards a unified view of state and behavior. *Information Systems* **28**, 5. http://AOR.rezearch.info.

Winikoff M, Padgham L and Harland J 2001 Simplifying the development of intelligent agents. In *AI2001: Advances in Artificial Intelligence. 14th Australian Joint Conference on Artificial Intelligence*, Adelaide, December 2001, pp. 557–568.

Wooldridge M 2002 *An Introduction to MultiAgent Systems*. John Wiley & Sons, Chichester, UK. ISBN 0 47149691X, http://www.csc.liv.ac.uk/~mjw/pubs/imas/.

Wooldridge M and Jennings NR 1995 Intelligent agents: theory and practice. *The Knowledge Engineering Review* **10**(2), 115–152.

Wooldridge M, Jennings N and Kinny D 2000 The Gaia methodology for agent-oriented analysis and design. *Journal of Autonomous Agents and Multi-Agent Systems* **3**(3), 285–312.

Index

Developing Intelligent Agent Systems   L. Padgham & M. Winikoff
© 2004 John Wiley & Sons, Ltd   ISBN: 0-470-86120-7 (HB)